EXCEL

2021

The All-In-One Beginner To Expert Excel Guide.

Learn The Excel Basics In 30 Minutes, Discover Formulas, Functions, Tips, And Tricks To Become a PRO.

+ Tutorials & Practical Examples

JASON MILLER

TABLE OF CONTENTS

INTRODUCTION

Have you ever wondered what excel could do for you? You will learn from this book how to use it and what you can use to help with your work. This book will discuss; creating an excel document, saving an excel file, adding items into the document, and closing and starting a new file.

Why should you use Microsoft excel? The new features of Excel 2011 help you do your work better. By learning the new features, you will be able to do more with the existing features. You can change all your existing sheets to make it easier to manage your data. You can learn about using formulas with the new features of Excel 2021, like conditional formatting and understanding where formulas are used in an excel sheet.

Office 2021 is a replacement and an upgrade version of Office 2019 and Office 2021. Office 2021 is a single payment or non-subscription version of Microsoft Office that comes along with many added features and upgrades, specifically created for consumers and small businesses.

Office 2021 is also designed for consumers who don't want to subscribe to the cloud-powered Microsoft 2021 variants.

Office 2021 contains the same applications as previous versions of Office, such as Word, Excel, PowerPoint, Outlook, OneNote, and depending on the plan or bundled subscribed to. You may get to use other applications and services such as Publisher, Planner, OneDrive, Exchange, SharePoint, Access, Skype, Yammer, and Microsoft Teams.

Office 2021 comes in two versions; consumer Office 2021 and Office LTSC (Long Term Servicing Channel) for commercial customers. Office LTSC includes enhanced features such as dark mode supports for visuals, accessibility improvements, and performance improvements on Word, Excel, and PowerPoint. Consumer Office 2021 also has similar features to Office LTSC for commercial users.

These versions of Office 2021 are both compatible with Windows and Mac.

This book will assist you in learning a new ability, namely a complete comprehension of Microsoft Excel, but you will also benefit from extensive practice. So, essentially you will be studying and practicing various ideas, functions, and formulas. You will have created your project or tiny app that you will use later throughout your personal, professional, or school life. You will have an amazing command of Excel, and you will also gain a lot of knowledge from the process so that you will not have to go through a long learning curve.

Herein is a guide to the basics of the program, plus the various types of problems that you could be encountering while using Excel. Once you have gone through this book, you should be able to solve all types of problems encountered in finance, economics, business, and accounting.

Contents include: Getting Started with Excel; General Editing Techniques; Understanding Worksheets; Selecting Cells; Moving Around in a Worksheet; Understanding the Ribbon Menu; Selecting Rows and Columns; Selecting Cells by Using Keyboard Shortcuts; Working with Names in Excel; Advanced Formatting Techniques; Applying Cell Styles; Checks Your Math for You; Printing your worksheets in Microsoft Word! All Wrapped up.

HISTORY OF EXCEL

The history of Microsoft Excel began in 1983 when Microsoft released its first spreadsheet program, Microsoft Multiplan. Excel was created to provide spreadsheets on the Macintosh operating system. Excel 2.0 was released for Mac in 1985 and came with about 100 built-in functions, multi-user support, and a reworked modeless interface that addressed many of the shortcomings of its predecessor's interface. This made it popular among lovers of the Mac at the time because you could use it with other applications inside one container called HyperCard to create documents that are interactive.

With each new version, Microsoft added more functions and introduced new concepts to make it easier for the user to organize and understand the content. This, however, had limitations as it was not totally compatible with Windows platforms at first.

Version 3.0 of Excel was released in 1990 during the Mac OS 9 era. It included an icon-based interface that was even more friendly than the previous one, making it especially useful for non-Microsoft Windows users or those whose first language is not English. However, Excel 3 was not available on Windows operating systems at the time, though they were able to open files created by Macintosh computers running Excel 3.

Excel for Windows 95 was released in 1991, picking up the GUI that was used on Mac OS 9. As with Mac OS 9, there were still significant limitations on how it could be used with non-Windows operating systems at the time. This made it especially useful for users of Microsoft Windows because of its ease of use and compatibility with Microsoft's Windows operating system.

Excel for Windows 3.0 came out in 1992 and was a great improvement from Excel 2.0, as it allowed you to save files on both Mac and Windows systems, but also provided new features such as a range finder and better printing capabilities along with improved copy & paste operations.

Microsoft progressively improved Excel for Windows, eventually allowing it to read and write files in the same format as the Macintosh version. Excel 5.0 (for Windows) was released in 1993, followed by Excel 4.0 (for Mac) in 1994. They added auto filtering, object linking and embedding (OLE), and protected worksheets/workbooks with passwords.

When Microsoft began developing Office 95, now called Microsoft Office, they made sure to work on Excel with the same features as word and other office applications that were to be released that year. This allowed users to open and edit files in a similar way in all applications. This was a major change from previous versions of Excel, where only certain features were available.

This version of Excel was the first Windows application to support over 60 languages, including Japanese, French, German, Italian, and Spanish, giving it a wider appeal among non-English speaking users. More specific languages would be added from then onwards, such as Czech, Danish, Dutch, Finnish, Norwegian, Polish and Swedish.

Excel 2007 for Mac included a number of new functions and features, including a whole suite of visual basic functions, which made it easy for people to create macros in VBScript or Visual Basic for Applications (VBA). This had the advantage of making macros more accessible to non-programmers.

Excel 2007 for Windows also included some new features that have become popular among home users over time, including themes that change the visuals based on the files or documents they contained. It has since gained support for macros that can be used in VBA scripts.

Excel 2010 for Mac made several improvements to ease the users of the spreadsheet by providing a cleaner interface that was brighter, had smoother animations and graphics, easier to pick colors and shapes with Adobe Illustrator or Acrobat, making it easier to alter the size of objects on the screen. This also allowed users to use more advanced tools such as annotations on sheets while viewing them.

Excel 2013 made a number of improvements to its ability to create and share documents over the internet. This included new features such as a cloud storage service called SkyDrive, which allowed users to store files locally on their computer as well as access them from any other device with internet access by using Word, PowerPoint, Excel, or Outlook. It also included a feature called Office 365 which provides business-level solutions for companies based on Microsoft Office.

In 2016, Excel 2019 was released for both Mac and Windows computers providing a number of new features desired by customers over time, including 3D maps, support for connecting with Power BI allowing users to create charts with data from sources such as Microsoft Analysis Services and PowerPivot databases. New features include control changes within charts, more options for data labels, and conditional formatting.

One of the most notable improvements to Excel was the introduction of the pivot table, which allows users to quickly and easily manipulate complex data. It became very popular as it allowed users to summarize data that would have otherwise been too difficult or time-consuming to handle by hand. Now that this is included in Excel, it can be found in other Microsoft Office applications, including Word and Access.

Other features released in this version are support for formatting within formulas, formatting text in categories, hiding or showing columns or rows, creating outlines, adding notes to cells, and creating custom menus.

To this day, the latest Microsoft excel 2021 remains one of the most commonly used applications by many businesses around the world.

Chart of Excel Versions

#	Name	Released
1	Version 1	1985
2	Excel 2	1987
3	Excel 3	1990
4	Excel 4	1992

5	Excel 5	1993
6	Excel 95	1995
7	Excel 97	1997
8	Excel 2000	1999
9	Excel 2002	2001
10	Microsoft Office Excel 2003	2003
11	Microsoft Office Excel 2007	2007
12	Microsoft Office Excel 2010	2010
13	Microsoft Excel 2013	2013
14	Microsoft Excel 2016	2016
15	Microsoft Excel 2019	2019
16.	Microsoft Excel 2021	2021

Versions of MS Excel

Excel has been through a lot of changes over the years to become the strong spreadsheet app that we understand and enjoy today.

We wanted to create a brief overview of the metamorphoses that Microsoft Excel has passed through over the years, motivated by our passion for the program. So, how about we take a step back in time?

Now we learn each past and latest MS Excel versions;

Version 1

Excel was first launched exclusively for Mac. Many Excel users are unaware of this, and it can seem unusual. Microsoft had already attempted to create a spreadsheet program named Multiplan in 1982, but it was unsuccessful. Until 2016, Excel versions for various operating systems were known by different names.

Excel 2

To equal to the Mac version, the very first Microsoft Excel variant for Windows was called "2." It was a port of the Mac "Excel 2," which had a run-time version of Windows.

Excel 3

Toolbars, drawing capability, highlighting, add-on support, 3D maps, and several other new features were included in this version.

Excel 4

Excel 4 Version was the first "popular" version of Excel. Many usability enhancements were added, including AutoFill, which was first included in this version.

Excel 5

Excel 5 was a significant improvement. This included multi-worksheet workbooks as well as VBA and Macros support. Excel became more resistant to macro virus attacks as a result of these latest additions, which would continue to be a concern till the 2007 version.

Excel 95

That was the first big 32-bit edition of Excel, and it was known as Excel 95. Excel 5 had a 32-bit variant as well, although it was not commonly adopted due to distribution bugs. Excel 95 is somewhat close to Excel 5 in terms of features. You may even be asking why Excel 6 isn't accessible. After Excel 7, all Microsoft Office programs have used the same version number, and the version numbering system has been modified.

Excel 97

This update included a modern VBA developer interface, data validation, User Forms, and many more. Can you remember Clippy, the obnoxious Office Assistant? He was also a part of this edition.

Excel 2000

HTML as a local file format, a "self-repair" capability, pivot charts, and upgraded clipboard and modeless user forms are among the new functionality.

Excel 2002

This is the very first time Excel has been used in Office XP. The large list of new additions didn't contribute much to the typical user's experience. The latest mechanism that helps you to restore your work if Excel crashes proved to be one of the most important features. This edition also has a helpful function named product activation technology which is also known as copy protection, which limits the usage of the software to one machine at a time. Until considering whether or not to update, think of the consequences.

Microsoft Office Excel 2003

Improved XML support, a fresh "list range" function, corrected statistical functions, and Smart Tag enhancements were among the new features in this version. The majority of users would not think the data update worthwhile.

Microsoft Office Excel 2007

Excel began to change in this Windows version. The Ribbon interface was introduced, as well as a change in the file format from .xls to now-famous .xlsx & .xlsm. This update improved Excel's stability (remember the issues with macro problems in previous versions?) and provided for more row data storage (above one million). The charting functions have also been significantly enhanced. To the delight of some and the chagrin of others, the upgrade included the elimination of Clippy from Microsoft Excel.

Microsoft Office Excel 2010

Pivot table slicers, Sparkline graphics, a 64-bit version, and an updated Solver were among the new features of this Excel upgrade. You may be wondering why Microsoft missed version 13 in favor of 14, but it's because the number 13 is deemed unlucky.

Microsoft Office Excel 2013

Over 50 new features were added in this version, and also a single-document interface suggested pivot tables and charts, as well as new charting improvements.

Microsoft Office Excel 2016

Despite the fact that these are separate versions of the program, Excel for Windows and Mac are now known as the same thing. If you have a Microsoft Office 365 account, you will get exclusive Excel Internet upgrades that can drastically improve the user experience. Older versions and those purchased at a store are also at a disadvantage. Histograms (to visualize the frequency of data), Pareto charts (to display data trends), and PowerPivot (which has its own language and allows for the importing of high levels of data) are among the latest features in this edition.

Microsoft Office Excel 2019

This is the successor of its former version, and it contains a wide assortment of new and improved features and capabilities.

Microsoft Office Excel 2021

As of this writing, this is the most recent edition of Microsoft Excel. It, of course, has many of the functionality used in previous versions of Excel, as well as some new ones. The updated charts, which provide a new spin to data presentation, are one of the most noticeable new additions. Funnel charts and Map charts are two examples of new data presentation charts that render the data appear perfect. In addition, you have the choice of including 3D graphics in your workbooks.

If you have an older version of Excel, it would probably work for newer files if you use the compatibility option.

Please remember the older models have even fewer functions, which is obvious if you've been paying attention to the table. Any of them might not be compliant with newer operating systems, so that's a good idea to test out various models and see how the similar file looks in each.

EXCEL 2021

Excel 2021 is a spreadsheet computer program that lets you create and organize data either in a printed format or in an electronic form. A spreadsheet also lets you take numerical data and combine it with other information such as text, pictures, charts, symbols, and graphics to create meaningful graphic representations of data. Excel contains features that allow you to carry out several operations such as calculation, graph tools, pivot tables, macro programming, and many more.

Microsoft has launched excel 2021 with some new features and functionalities, which make it better than all other versions. Some of these features are the ability to work with cells while editing, the ability to export images from various applications, color coding of various information in excel, etc. These functions will definitely help you quickly perform various tasks without any hassle.

Excel 2021 uses cloud storage to save its file and can be accessed from a web browser on the computer system.

MS Excel 2021 Features

Excel and traditional Excel have some features in common. Right now, I will be listing out some features that make Excel different from traditional Excel.

Online Subscription

Excel 2021 is the subscription-based version of Excel designed to regularly release updates and features that will enhance the productivity of its users. The subscription payment can be done monthly, semi-annually, or annually.

Custom Visuals

One of the features available in Excel 2021 is custom visuals such as bullet charts, speedometer, and word cloud, which were available only available in Power B1.

Custom Functions

This feature allows you to create custom functions by using JavaScript, which permits for better interconnection.

Full SVG Graphics

Excel 2021 comes with SVG graphics support and 500 built-in icons, which look great on infographics and dashboards.

3D Models with Full Rotation

Excel 2021 has many 3D models that are for free on the internet, with extensions such as. fbx., obj., ply., stl., and gbl.

XLOOKUP Function

Another feature in Excel 2021 is the XLOKUP function. This function allows you to find the value that is located within a spreadsheet range or table.

More Images, Icons, Backgrounds, and Templates

Excel 2021 comes thousands of new designs such as images, icons. Backgrounds and templates.

Ideas

Another feature in Excel 2021 is the Idea function. The Idea function offers help on how to express data or put them into visualization.

Black Theme

The black theme in Excel 2021 makes late-night work editing with ease.

Split Columns to Rows

This is a new feature in Power Query where each delimiter generates a new row.

Funnel Chart

This is a chart type that comes in handy for illustrating a sales funnel.

Co-authoring Features

This feature in EXCEL 2021 allows two or more users to edit a workbook when stored on OneDrive or SharePoint simultaneously.

Importance of Using EXCEL 2021

Now let's talk about the importance of using Excel 2021 compared to traditional Excel.

- **Preparation of Financial data:** One of the reasons to use Excel 2021 is that it allows you to prepare financial data such as budgets, account balance information, taxes, payrolls, receipts, and a lot more.

- **Mathematical Formulas:** With Excel 2021, you can solve complex mathematical problems by making use of the mathematical formulas in Excel.

- **Online Storage and Access:** Excel 2021, which is a part of Office 2021, allows you to access their files online without the need to move around with their computers. In a nutshell, you can access your files anytime and anywhere using any device compatible with the use of Excel 2021

- **Easy and Effective Comparison:** With Excel 2021, you can analyze a large amount of data that can be used to get trends and patterns that can influence or affect decisions.

- **Co-authoring:** Excel 2021 allows you to work on the spreadsheet at the same time with other users.

- **Improved Security:** In contrast to the traditional Excel, Excel 2021 offers an advanced security system to the files on it. This denies intruders access to the files by either using a password using the Visual Basic Programming or directly within the Excel files.

- **Creating Forms:** With Excel 2021, you can create form templates that can be used for handling inventories, performance, evaluation, questionnaires, and reviews.

GETTING STARTED WITH EXCEL

Excel is a useful program, but it can sometimes be perplexing. This is why this book has all of the knowledge you'll need to get started using Excel. It will guide you through the whole procedure, from opening a spreadsheet to inputting and manipulating data in preparation for storing and sharing. To proceed, make sure you have Excel 365 installed on your computer or mobile device.

What Is the Procedure for Obtaining the Microsoft Excel Application

You will not be capable of using Microsoft Excel unless you have previously downloaded it on your device or can access it from anywhere. Microsoft Excel may be obtained in a variety of ways. It should be purchased from a computer hardware store that also sells software. Excel is a piece of software that comes with the Microsoft Office suite. The license key may also be downloaded from the Microsoft website.

Office Software Purchase and Installation

Many of the Microsoft Corporation's important products are included in the Microsoft Office program. You may access various Microsoft applications, including Excel, after purchasing and installing this software on your smartphone. The office pack is available for purchase on the Office's website and other online and offline retail locations.

Before you install Excel 2021 on the device, it's a great way to learn that Windows 10 is the least Windows update that it will support. If your computer runs Windows 7 or 8, you might have to upgrade to Windows 10 before installing it.

If you don't currently have an Office pack on your phone, you might consider upgrading to the 365 upgrades, which provide several useful features. Several sites provide where you can get an Office for your computer and update it. If you go to Microsoft website, for example, Microsoft recommends that you purchase the Microsoft 365 software for your computer.

As a result, the app includes many new functional improvements to the package's components. On the opposite side of the Amazon website, you will be sent to the website's homepage. Choose software as your search option from the drop-down menu at the upper left-hand side of the site's search box; whatever computer equipment vendor store near you may sell you any Office software you choose. The vendor may even start the installation and activation procedure for the software bundle.

Obtaining Microsoft Office for Free

Everyone will get about a one-month of the free trial of Microsoft 2021(365). You must input the credit card because you will be charged around $100 for a one-year membership to the family of Microsoft 365 if you do not cancel before the end of the month (formerly referred to as Office 365 Home).

In case you don't need the full suite of the Microsoft 365 apps, the good news is that many of them, including Word, Excel, OneDrive, PowerPoint, Calendar, Outlook, & Skype, are available for free online. How to have your hands on them is described below:

Go to Microsoft Website

Log in to the Microsoft account, then create a new account (or create 1 for free). Whether you use Windows, Xbox, or Skype, you have a Microsoft account.

Choose the program you wish to use & save the work to the cloud via OneDrive.

Getting Into/Out of Office 365

Your Microsoft account will allow you to log into Office 365 easily. Creating a new account, though, is simple. To get started, follow the steps below:

1. To get started, go to the Microsoft website.

2. If you don't already have one, enter your email address & click "Next" or "Create One" if you don't.

3. Enter in your password & choose "Keep me logged in" to avoid having to type in each time you log in.

Sign in

Use your Microsoft account.
What's this?

| Email or phone |
| Password |

☐ Keep me signed in

Sign in

No account? Create one!

4. You've arrived at the Office home page. On this page, you can easily choose the Office 365 bundle for 2021 to use. Alternatively, you may click the "Install Office" option to download the full Office 365 suite to your computer, then subscribe by clicking the "Buy Office" button. You may also get started with the free edition.

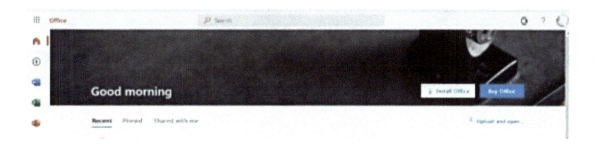

What Is the Best Way to Open Microsoft Excel

On the left side of the Office 365 main tab is the Navigation bar. You have the option of launching each app individually. Go to the app launcher and pick the Excel icon. You'll then be sent to a new page where you may create a new worksheet, access recent data, and utilize templates. Click New blank workbook to start a new page in Excel.

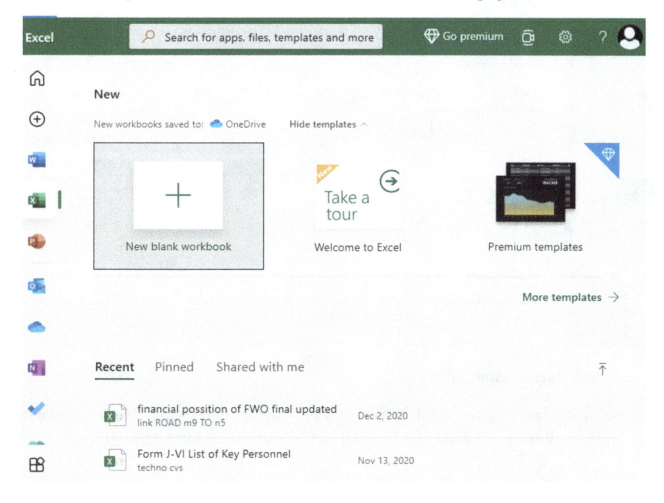

Choose a file from the Recent to view a previous work on the Worksheet. Click upload and open to access Excel documents on your computer. Click more templates to get a list of available templates. You should also keep in mind that the data is stored in the Microsoft cloud by default. You may, however, download any material to your computer system if required.

Excel's Start Page

When you open Excel or click on Excel, the Start Screen displays on. On this screen, the right and left panels are divided. The left panel becomes green with the home icon chosen, with New Open items at the top, and Account, Feedback, & settings options at the bottom. A chart of thumbnails with a choice of templates to select from is shown on the right side. To establish a new blank Excel workbook file, click the blank workbook button on the right side.

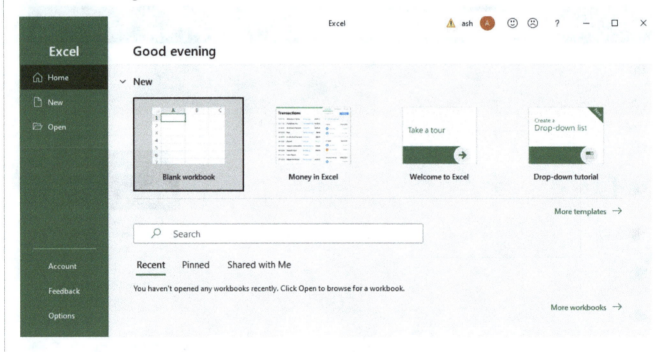

Excel Workplace Customization

The Microsoft Excel framework depicts how users interact with the MS Excel program or an excel application. The following describes the elements of Microsoft Excel. The

Microsoft Excel GUI makes MS Excel workbook operations and procedures more efficient.

Thanks to the Live Preview feature, the frame's formatting has improved. To view the format in the browser, move the mouse pointer over the instruction.

A couple of the settings listed below may be used to change the way the GUI works.

Creating an Excel Template

- Start by opening Microsoft Excel and making a new blank workbook.

- Save the workbook with a specific file name in the necessary folder.

- Below are some more suggestions and actions to take. There are a few Excel workbook basics that must be changed.

Font Type and Size

Select your number, alignment, and font type choices from the Font list at the beginning of the same Worksheet, then highlight the sections of each Worksheet you want to change.

Column Width/Layout

Typically, you choose different column widths, then select the columns or even the whole sheet, and afterward modify the column width.

In Print Settings, choose one or more worksheets, then go to Page Layout, then Page Setting to customize print settings, including headers & footers, page orientation & margins, and a variety of additional print layout choices.

Gridlines

Would you like darker gridlines for each Worksheet? The black grid lines or borders are visible but do not print. Go to file> Options > Advanced to alter the gridline color. Then choose Display options for the current Worksheet from the drop-down menu, followed by the workbook title from the drop-down menu. Finally, choose a gridline color from the Display gridlines menu.

Worksheet Count

You may add and remove worksheets, as well as rename and color sheet tabs.

- **Note:** When you add a new worksheet, the layout and style of your customized default workbook will be restored. You'll want to attach extra worksheets to the real workbook to separate an optional or primary worksheet that you may duplicate if desired.

Choosing a Color Scheme

Go to the Excel ribbon and choose the File Options option to change the color theme for your Excel sheet. The browser will start as you go through the instructions below.

- On the other hand, the left-hand side's general tab will be chosen.

- Look for a color scheme under General options for Excel dealing.

- From the color scheme drop-down column, choose a suitable color. Choose the OK tab.

Formula Parameters

This option may help you explain how Excel works with formulae. You'll use it to make choices like utilizing a fully automated layout, changing the cell reference style to use the column to row numbers, and more. Select the checkbox if you wish to start a function. If you wish to deactivate an option, delete the checkbox icon. The Options window in the Method tab's left-hand column has this option.

Proofing Options

As an alternative, it manipulates the supplied text in Excel. When searching for erroneous spellings, tools such as dictionary vocabulary, dictionary suggestions, and so on may be used. This option is under the check tab on the left-hand side of the dialogue box.

Getting Familiar with Excel Screen Interface

Let's learn and find out about Excel Interface, which includes the Start Screen, Ribbon Interface, and how to customize the ribbons on Excel's interface.

Excel's Start Screen

When you open an Excel application for the first time, the first thing that pops up is the start screen, which is divided into two parts. The Left Navigation Pane and the Right Navigation Pane.

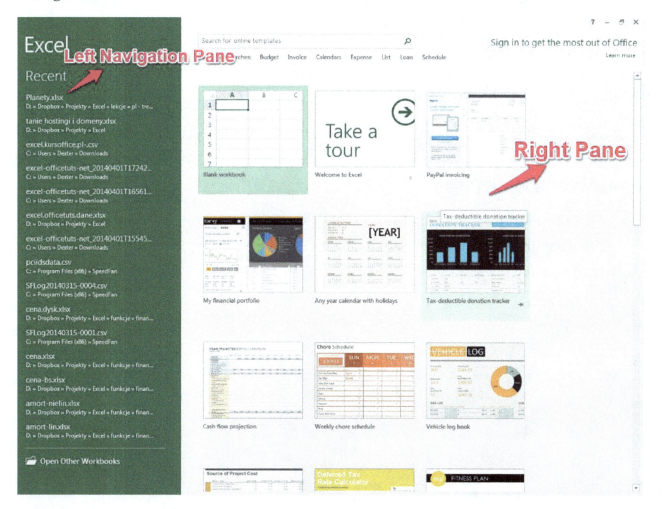

The Left Navigation Pane

The Left Navigation pane comprises a list of recently opened Excel files and a link which is "Open Other Workbooks." When you click on "Open Other Workbooks," this takes you to the backstage view of Excel, where you can access options such as New, Open, Save, Save As, etc.

The Right Pane

The Right Pane displays a list of thumbnails that includes templates that can be used to create a new workbook. To view more templates to create a new workbook, In the New link on the right side of the Home screen, click on Find more. To open a new blank Excel workbook, you can click on Blank workbook.

Excel's Workbook User Interface

From the Excel Home Screen, you open a new, blank workbook by clicking on the New Workbook thumbnail. When a new, blank workbook is opened, the following options are displayed on the user interface.

- **File Menu Button:** The File Button takes you to the Backstage View of Excel, and this contains several options such as New, Open, Save As, Print, etc., to work with the Excel file.

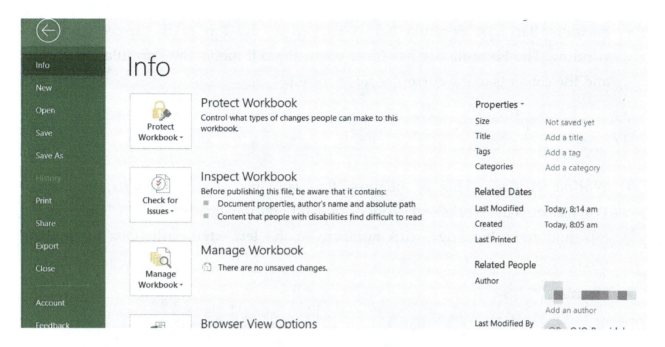

- **Quick Access Toolbar:** This tool is located above the Excel ribbon, and by default, it contains Commands such as Save, Undo, and Redo. This Quick Access Toolbar can be customized by adding any other commonly used command to it by clicking on the Customize Quick Access Toolbar button beside the Quick Access Toolbox button.

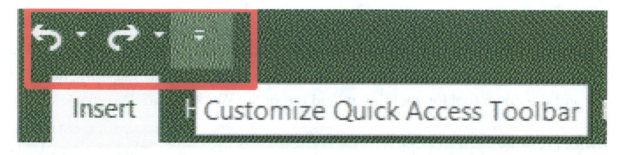

- **Ribbon:** This contains most of the commonly used commands in Excel. They are displayed on the Excel interface in tabs ranging from the Home tab to the View tab.

File	Insert	Home	Draw	Page Layout	Formulas	Data	Review	View	Developer	Help

- **Formula Bar:** The Formula bar is located at the top of the Excel worksheet window. The Formula bar has three parts; the cell name, the Formula bar button, and the contents of the currently selected cell.

- **Worksheet Area:** This is the area that contains all the cells in the current Worksheet. The Worksheet is identified by column headings with letters at the top, and rows headings with numbers at the left edge, with tabs for making selections.

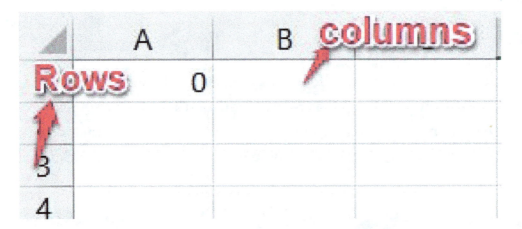

- **Status Bar:** The Status bar keeps you abreast of the current model of the Excel worksheet you are engaged with. The Status bar also contains the worksheet views and the Zoom tool for zooming in and out of the Worksheet.

- **Windows Controls:** The Window controls are used to control the main Excel window. The Window controls contain three buttons; maximizing the window, restoring the window, and closing the window.

- **Ribbon Display Options:** The Ribbon Display Options button is located at the top of the Excel window, and when clicked on, the three options are displayed; Auto-hide Ribbon, Show Tabs, and Show Tabs and Commands.

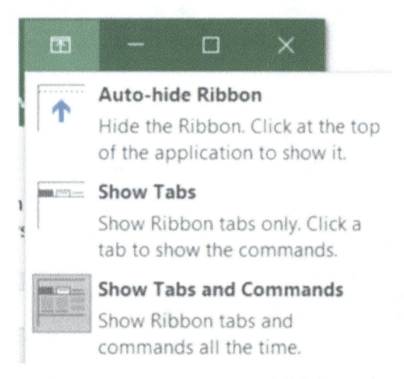

- **Horizontal Scrollbar and Vertical Scrollbar:** The Horizontal scrollbar and Vertical scrollbar are used to scroll the content in the Worksheet horizontally or vertically.

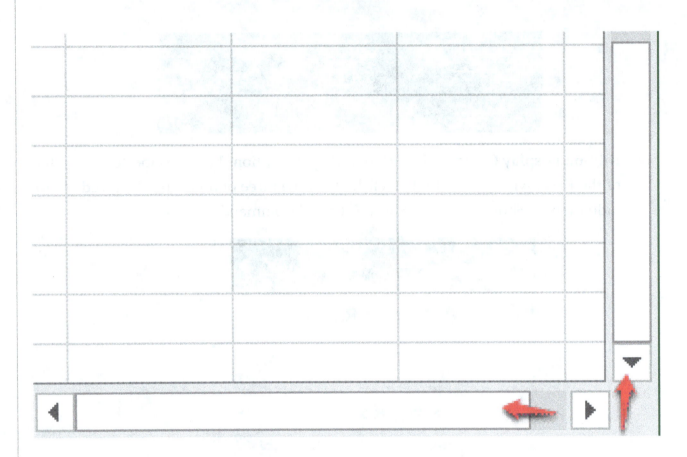

Navigating Through the Excel Ribbon

The Excel ribbon is a row of tabs, buttons, and icons located at the top of the Excel window. These tabs, icons, and buttons are categorized based on their functions or categories.

Components of Excel Ribbons

The Excel ribbons are divided into four components; Tabs, Groups, Buttons, and Dialog Box launcher.

Tabs

Tabs are a group of commonly used commands brought and displayed to perform an essential task. The following are the tabs in Excel.

- **File Tab:** This is the first tab in Excel which is used to open the Excel Backstage View. The Excel Backstage View has several options for customizing, editing, and managing Excel files.

- **Home Tab:** The Home tab contains commands that are commonly used in Excel, and some of these commands are copy, paste, format, find, replace, etc. The Home tab is arranged into the following groups; Clipboard, Font, Alignment, Number, Styles, Cell, and Editing.

- **Insert Tab:** The Insert tab contains objects or elements that can be inserted into the Worksheet. The elements include graphics, pivot tables, charts, hyperlinks, shapes, 3D models, pictures, etc. The Insert table is arranged into the following groups; Table, Illustration, Apps, Charts, Reports, Sparkline, Filter, links, Text, and Symbols.

- **Page Layout:** The Page Layout tab contains options for Excel page setup and print. The Page Layout tab is arranged in the following group; Themes, Page Setup, Scale to Fit, Sheet Options, and Arrange.

- **Formulas Tab:** This tab contains options for adding formulas and functions in a worksheet and troubleshooting the functions for errors. The Function tab is arranged in the following group; Function Library, Defined Names, Formula Auditing, and Calculation.

- **Data Tab:** The Data tab contains options for filtering, sorting, and manipulating data. The Data tab is arranged in the following groups; Get External Data, Connections, Sort & Filter, Data Tools, and outline.

- **Review Tab:** The Review tab contains options for spell checking, thesaurus, sharing, protecting, and tracking changes in the Worksheet. The Review tab is arranged in the following groups; Proofing, Language, Comments, and Changes.

- **View Tab:** The View tab contains options for changing the display of the Worksheet and its contents. The view tab is arranged in the following groups; Workbook View, Show, Zoom, Windows, and Macros groups.

- **Developer Tab:** The Developer tab contains options for creating, playing, and editing macros. It can also be used to import and map XML files. The Developer tab is arranged in the following group; Code, Add-ins, Controls, and XML.

- **Help Tab:** The Help tab is where you get online help and training and feedback on Excel.

Groups

The groups contain related commands buttons which are arranged into subtasks. Each contains buttons, sub-menu, and dialog launchers.

Command Buttons

These are tools in the group that are used to execute an action in the Worksheet. The

command buttons in the tab are organized into mini-toolbars.

Dialog Box Launcher

The Dialog Box Launcher is located at the right bottom corner of each group. When you click on it, the Dialog Box launcher opens a dialog box that displays additional options that can be selected from.

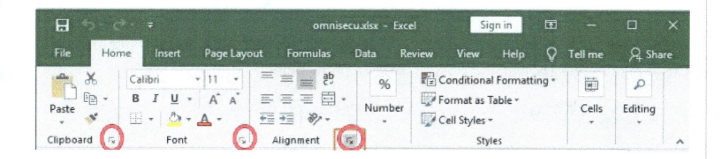

Customizing the Quick Access Toolbar

The Quick Access Toolbar is located above Excel Ribbon, and by default, contains the following four buttons:

- **AutoSave:** This option automatically saves all the future edits made in the Worksheet.

- **Save:** This option allows you to manually save the changes made to the Worksheet you are currently working on.

- **Undo:** This option undoes the last editing action made on the Worksheet you are currently working on.

- **Redo:** This option repeats the previous editing action recently removed using the Undo bottom.

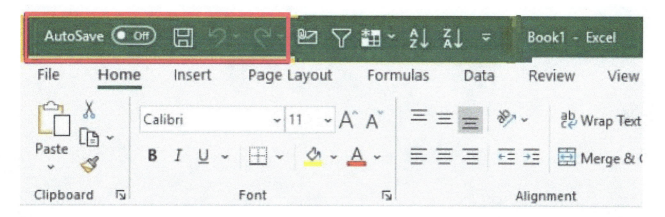

Ordinarily, you can append commands for Quick Access Toolbar by clicking on the Customize Access Toolbar button located beside the Quick Access Toolbox button.

By default, the Quick Access Toolbar is located at the top left corner of the Excel application, and it can also be moved under the Ribbon area by clicking on the Customize Access Toolbar and then clicking on Show Below.

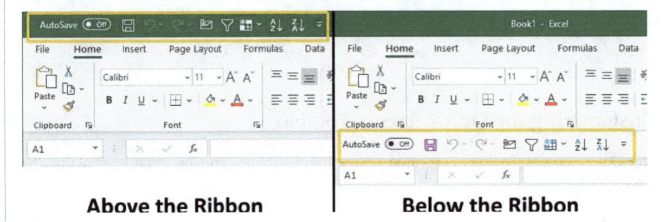

Above the Ribbon   **Below the Ribbon**

FUNDAMENTALS

Cell Basics

A Worksheet's basic building blocks are cells. To measure, evaluate, and manage data in Excel, you'll need to understand the basics of cells and cell content.

Understanding Cells

Thousands of rectangles referred to as cells make up each Worksheet. The letters A, B, and C identify the columns, while the numbers identify the rows (1, 2, 3). Based on its column and row, each cell has its name or cell address. The selected cell in the example below intersects row 5 and column C, so its cell address is C5.

To Select a Cell

Select a cell by clicking on it. When you select a cell, you'll notice that the cell's borders become bold, and the row headings and cell's column become highlighted.

Until you click on an area in the Worksheet, the selected cell will remain selected.

You can also use your keyboard's arrow keys to move through your Worksheet and choose a cell.

To choose multiple cells, use your mouse to click and drag until all the adjacent cells you require are highlighted.

Cell Content

Each cell may contain different types of content, including its formatting, text, formulas, functions, and comments.

Formatting Attributes

Formatting attributes in cells will modify how numbers, dates, and letters are displayed. For example, percentages can be represented as 0.20 or 20%. You can also change the background color or text of a cell.

Text

Numbers, letters, and dates can all be inserted in cells.

Formulas and Functions

A cell can contain functions and formulas for calculating cell values. =SUM (cell 1, cell 2...), for example, is a formula that adds the values in multiple cells.

Comments

Multiple reviewer comments can be contained in cells.

To Insert Content

- Select a cell by clicking on it.

- Using your keyboard, type content into the selected cell. The content is displayed both in the cell and the formula bar. You can also enter or edit cell content in the formula bar.

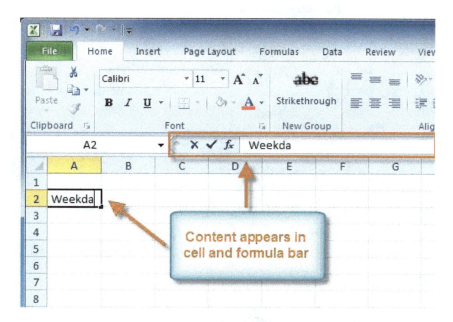

- To delete content within cells, choose the cells you want to delete from the spreadsheet or those that you do not require. On the Ribbon, choose the Clear command. A dialogue box will appear. Then select Clear Contents.

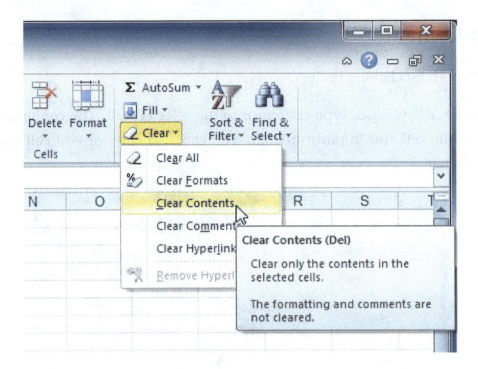

You can also remove content from a cell by pressing the Backspace key on your keyboard or delete data from numerous cells by pressing the Delete key.

To Delete Cells

- Choose the cells you want to delete. From the Ribbon, select the Delete command.

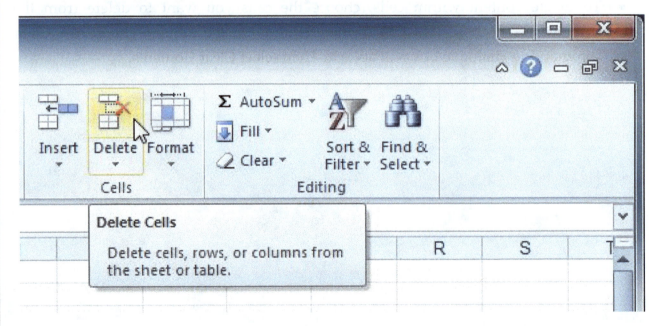

The difference between deleting a cell's content and deleting the cell itself is important. When you delete a cell, the cells underneath it will automatically shift up to take their place.

To Copy and Paste Cell Data

You can save effort and time by copying and pasting content from your spreadsheet into other cells in Excel.

- Make a selection of the cells you want to copy.

- Type Ctrl +C on your keyboard, or select the Copy option on the Home tab.

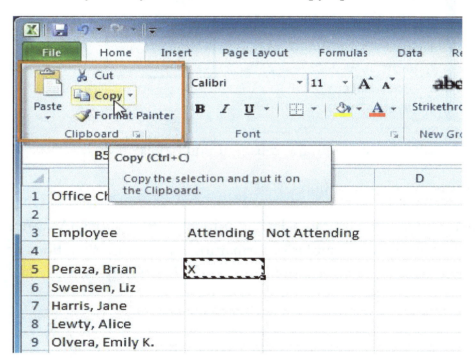

- Choose the cells you want to paste the content into.

- Type Ctrl +V on your keyboard or select the Paste option from the Home tab.

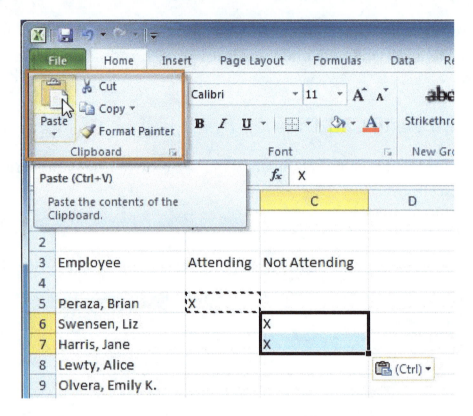

- The data will be pasted into the cells that you have chosen.

- To acquire more paste options.

From the Paste command, you can select additional Paste options. Advanced users who work with cells that contain formulas or formatting will find these options useful.

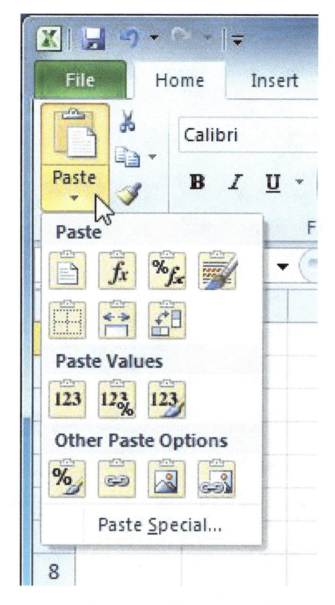

Instead of selecting commands from the Ribbon, you can easily access commands by right-clicking. Simply right-click the mouse and choose the cells you want to format. You'll see a drop-down menu with some commands that are also on the Ribbon.

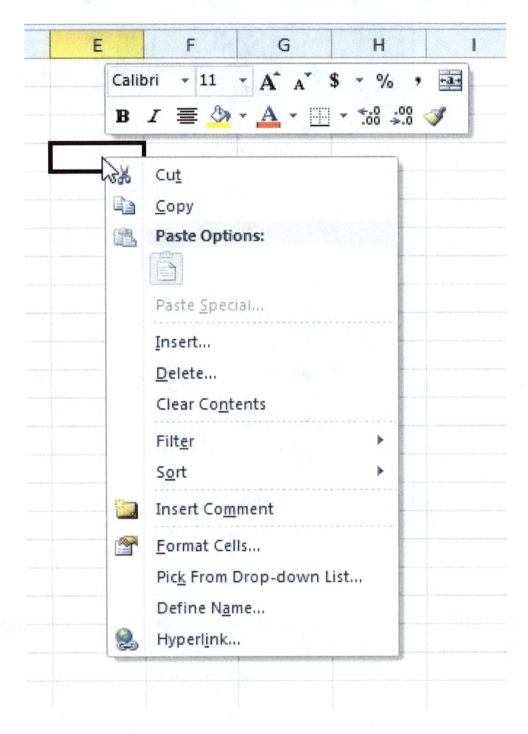

To Cut and Paste Cell Data

- Choose the cells you want to cut.

- Select the Cut option. The appearance of the chosen cells' borders will change.

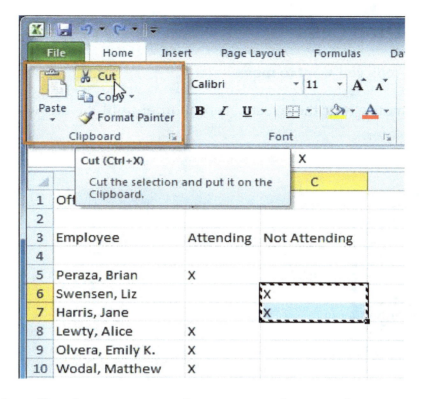

- Choose the cells where you want the content to be pasted.

- Select the Paste option. The cut content will be pasted into the selected cells after it is removed from the original cells.

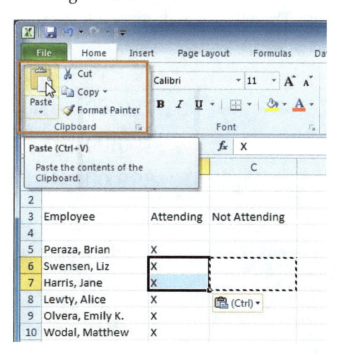

To Drag and Drop Cells

- Choose the cells you want to move.

- Hover the mouse over the selected cell's border until the arrow changes to a four-arrowed pointer.

	A	B	C	D	E
1	$1,799.00				
2	$200.00				
3	$99.00				
4					
5					
6					

- Drag the cells to their new location by clicking and dragging them.

- Release your mouse. The cells will be dropped into the specified place.

	A	B	C	D	E
1			$1,799.00		
2			$200.00		
3			$99.00		
4					
5					
6					

Formatting Cells

By default, all cell content has the same formatting, making a workbook with a lot of information difficult to read. Basic formatting will help you design the texture of your workbook, allowing you to highlight particular sections and making your content easier to understand and view.

To Change the Font Size

- Choose the cells you need to modify.

	A	B	C	D
1	FITNESS PROGRESS CHART			
2	Date	Weight	Chest	Waist
3	5/3/13	140	32	31
4	5/11/13	140	32	31

- Select the desired font size on the Home tab by clicking the drop-down arrow next to the Font Size command.

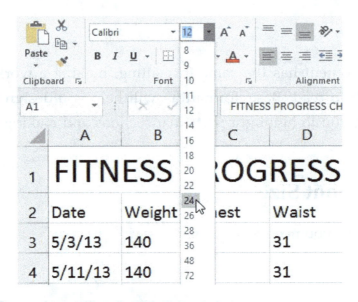

- The chosen font size will be applied to the text.

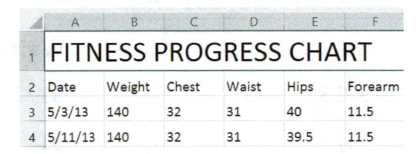

- You can also use your keyboard to enter a customized font size or use the Increase and Decrease Font Size controls.

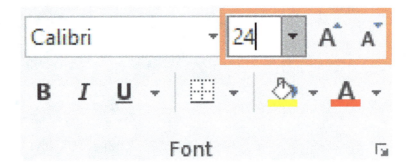

To Change the Font

Every new workbook's font is set to Calibri by default. However, Excel has a variety of fonts that you can use to modify your cell text.

- Choose the cells you want to update

	A	B	C	D
1	FITNESS⊕PROGRESS			
2	Date	Weight	Chest	Waist
3	5/3/13	140	32	31
4	5/11/13	140	32	31

- Select the desired font by clicking the drop-down arrow next to the Font command on the Home tab.

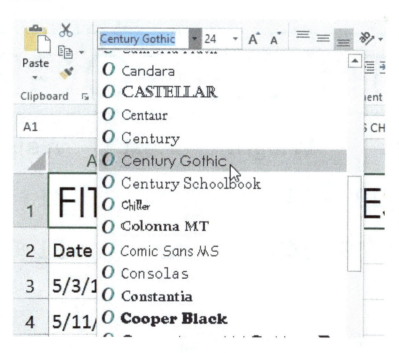

- The text will change to the font you've chosen.

	A	B	C	D
1	FITNESS PROGRE:			
2	Date	Weight	Chest	Waist
3	5/3/13	140	32	31
4	5/11/13	140	32	31

- When composing a workbook at the workplace, choose a font that is easy to read. Standard reading fonts include Times New Roman, Cambria, Calibri, and Arial.

	A	B	C	D	E
1	FITNESS PROGRESS				
2	Date	Weight	Chest	Waist	Hips
3	5/3/13	140	32	31	40
4	5/11/13	140	32	31	39.5

To Change the Font Color

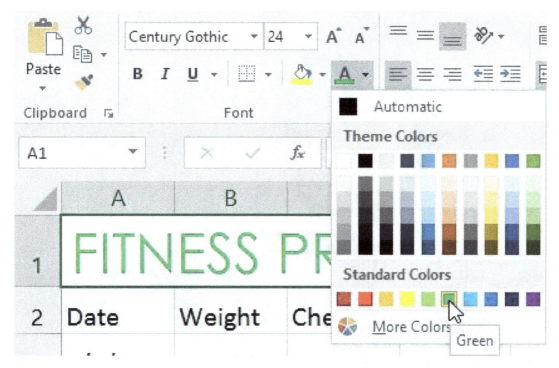

- Choose the cells you want to update.

- Select the desired font color by clicking the drop-down arrow next to the Font Color command on the Home tab.

- The text will change to the font color you've chosen.

	A	B	C	D	E
1	FITNESS PROGRESS				
2	Date	Weight	Chest	Waist	Hips
3	5/3/13	140	32	31	40
4	5/11/13	140	32	31	39.5

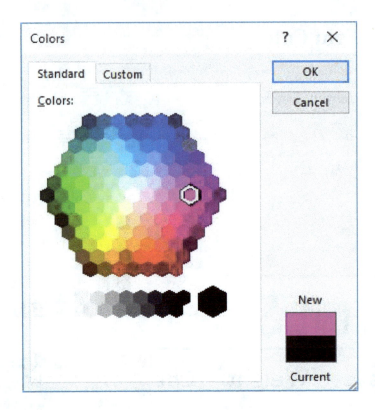

- To see more color options, go to the bottom of the menu and choose More Colors.

To Use the Italic, Bold, and Underline Commands

- Choose the cells you decide to modify.

- On the Home tab, select the Italic (I), Bold (B), or Underline (U) command. We use the bold command in the example below.

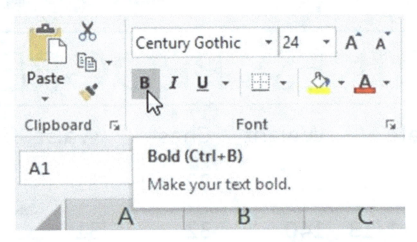

- The text will be styled in the chosen format.

	A	B	C	D	E
1	**FITNESS PROGRESS**				
2	Date	Weight	Chest	Waist	Hips
3	5/3/13	140	32	31	40
4	5/11/13	140	32	31	39.5

You can also make selected text italicized by pressing Ctrl +I, bold by pressing Ctrl +B and underlined by pressing Ctrl +U on your keyboard.

Fill colors and Cell borders

You can create defined and clear boundaries for different segments of your Worksheet using cell borders and fill colors. To further distinguish our header cells from the rest of the Worksheet, we'll add cell borders and fill color to them.

To Add a Fill Color

- Choose the cells you decide to modify.

	A	B	C	D	E	F	G	H	I
1	**FITNESS PROGRESS CHART**								
2	Date	Weight	Chest	Waist	Hips	Forearm	Estimated Lean Body	Estimated Body Fat	Estimated Body Fat %
3	5/3/13	140	32	31	40	11.5	103.8	36.2	0.259
4	5/11/13	140	32	31	39.5	11.5	103.9	36.1	0.258
5	5/19/13	139	32	31	39.5	11.5	103.2	35.8	0.258
6	5/26/13	138	31	30	39	11	103.4	35.6	0.256
7	6/1/13	138	31	30	39	11	103.4	35.6	0.256

- On the Home tab, select the fill color you want to use by clicking the drop-down arrow next to the Fill Color command.

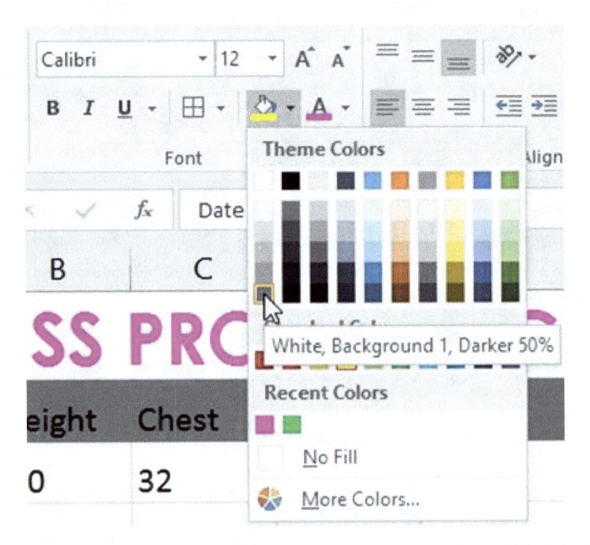

The fill color you choose will appear in the cells you select. We also modified the font color to white to make the dark fill color more readable.

	Date	Weight	Chest	Waist	Hips	Forearm	Estimated Lean Body	Estimated Body Fat	Estimated Body Fat %
1	**FITNESS PROGRESS CHART**								
3	5/3/13	140	32	31	40	11.5	103.8	36.2	0.259
4	5/11/13	140	32	31	39.5	11.5	103.9	36.1	0.258
5	5/19/13	139	32	31	39.5	11.5	103.2	35.8	0.258
6	5/26/13	138	31	30	39	11	103.4	35.6	0.256
7	6/1/13	138	31	30	39	11	103.4	35.6	0.256

To Add a Border

- Choose the cells you decide to modify.

	Date	Weight	Chest	Waist	Hips	Forearm	Estimated Lean Body	Estimated Body Fat	Estimated Body Fat %
1	FITNESS PROGRESS CHART								
3	5/3/13	140	32	31	40	11.5	103.8	36.2	0.259
4	5/11/13	140	32	31	39.5	11.5	103.9	36.1	0.258
5	5/19/13	139	32	31	39.5	11.5	103.2	35.8	0.258
6	5/26/13	138	31	30	39	11	103.4	35.6	0.256
7	6/1/13	138	31	30	39	11	103.4	35.6	0.256

- On the Home tab, select the border style you want to use by clicking the drop-down arrow next to the Borders command.

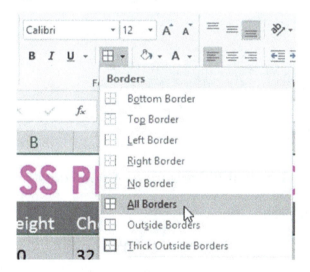

- The border style you choose will appear.

	Date	Weight	Chest	Waist	Hips	Forearm	Estimated Lean Body	Estimated Body Fat	Estimated Body Fat %
1	FITNESS PROGRESS CHART								
3	5/3/13	140	32	31	40	11.5	103.8	36.2	0.259
4	5/11/13	140	32	31	39.5	11.5	103.9	36.1	0.258
5	5/19/13	139	32	31	39.5	11.5	103.2	35.8	0.258
6	5/26/13	138	31	30	39	11	103.4	35.6	0.256
7	6/1/13	138	31	30	39	11	103.4	35.6	0.256

- With the Draw Borders tools at the bottom of the Menu drop-down, you can draw borders and adjust their line style and color.

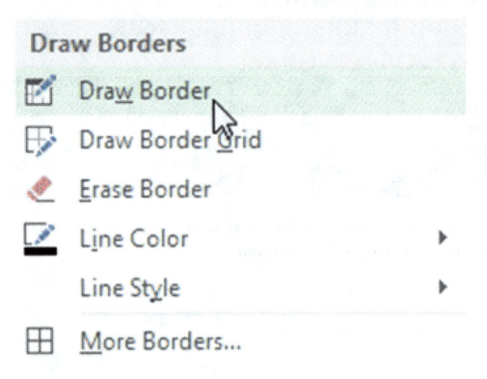

Align Text in a Cell

If you want to improve the visual presentation of your data by realigning text in a cell, follow these steps:

- Choose the cells in which you want the text to be aligned.

- Select one of the following alignment choices from the Home tab:

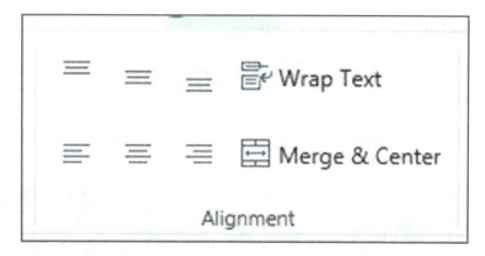

- Select Middle Align, Top Align, or Bottom Align to align text vertically.

- Select Align Text Right, Align Text Left, or Center to align text horizontally.

- If you have a large line of text, a part of the text will be hidden. Wrap Text can be used to correct this without changing the column width.

- Click Merge and Center to center text that spans several columns or rows.

Format Painter

You can find the Format Painter command on the Home tab. to copy formatting from one cell to another. The Format Painter will copy the formatting from the selected cell when you click it, which can then be pasted to any cells by clicking and dragging it over them.

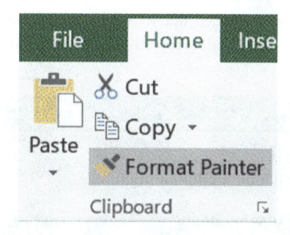

What Is the Quick Access Toolbar

The easy access toolbar in Excel (shown below) allows you to access the resources you use most in one touch.

By selecting the arrow icon on the hand side that is furthest to the right, you will configure what appears here.

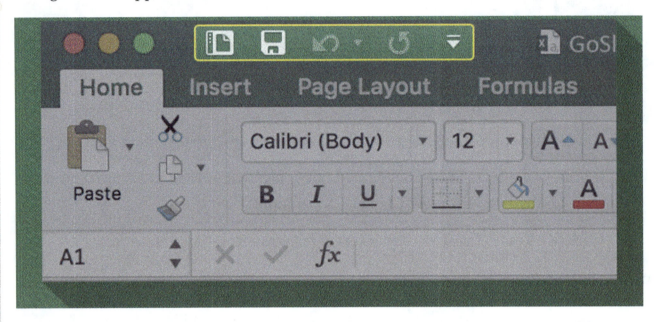

What Is the Ribbon

A Ribbon is a group of symbols at the top of the workbook that will assist you in performing various tasks. Consider the Ribbon as a hierarchy, with each tab (for example, Home, Insert & Page Layout) containing a set of commands (Filter, Sort, Copy, Paste).

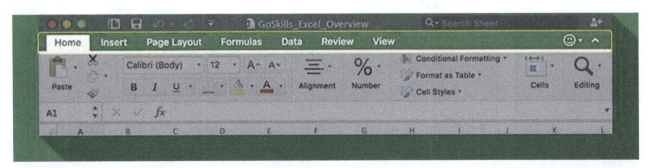

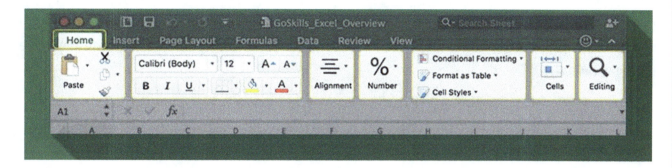

We will examine each of these in some detail.

What Are Tabs

Commands are found on the Ribbon (view screenshot). It's divided into clickable tabs, each of which contains a group of relevant commands.

It's necessary to remember that your Excel Ribbon might look different from mine. This may be for a variety of reasons.

Add-Ins

Add-ins expand the functionality of Microsoft Excel, necessitating the development of a new page.

Contextual Tabs

Contextual tabs, or tabs that show when you instantly do something unique in Excel, are also accessible.

A contextual tab is a kind of tab that only occurs when you choose a certain item, like a chart or graph. Contextual tabs include commands that are specific to the item you're focused on.

Customization

You may configure the Ribbon by inserting and/or deleting tabs to make the Excel function for you.

What are Worksheets and Workbooks

Excel is a (Microsoft Office) application that allows you to create workbooks and worksheets. Worksheets are sheets with columns, cells, and rows in them. A number, math formula, date, document, or Excel feature may be entered in each cell. Worksheets may also use a number of chart styles to represent specified information. A (workbook) is simply a collection of (worksheets). The consumer is provided with a workbook with three blank worksheets when the Excel sheet is first open. The first Worksheet appears, with three tabs numbered (Sheet1, Sheet2, and Sheet3) on the (below-left side), as seen in the (screenshots below). If a workbook has a huge number of worksheets, arrows appear to help the user (scroll) left and right to display worksheet tabs.

You will not have to uninstall the two useless worksheets if you're just using one; most people couldn't be bothered. In newer models of Excel, workbooks are saved with the ".xlsx" file extension. The ".xls" extension was included in previous releases. What is the maximum number of Excel worksheets that can be used in a single workbook? According to (Microsoft), the quantity is restricted only by your system (memory!) It's a good idea to combine worksheets that are very closely linked together, mainly if you're linked details between one Worksheet to the other. However, using the worksheet tabs to scroll back and forth may be difficult.

Viewing, Inserting, Deleting, and Renaming Worksheets

The name "Sheet1" isn't really (descriptive). You can access (rename), attach, and erase worksheets in the following ways in a workbook.

How to View a Worksheet

Tap on a "Worksheet's tab" to see it. Use the arrow to the left to move right or left, or right-click on each of the arrows and choose the required Worksheet from the collection that appears if the workbook browser is not wide to show all of the tabs due to multiple worksheet tabs or long worksheet titles.

How to Rename a Worksheet

Right-click on the spreadsheet link, choose Renamed from the menu bar and enter a new title. Alternatively, you can double-click the worksheet tab then enter a new title.

How to Insert a Worksheet

Simply click on the tiny tab (to the right of the last worksheet tab), as seen in the picture below, to add a worksheet to a workbook. The Worksheet may then be moved to another location if necessary.

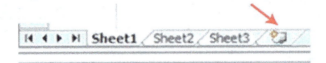

You may also place a new worksheet to the left of a current worksheet by right-clicking on the Worksheet tab directly to the right where you'd like the new Worksheet to go and selecting Insert from the Insert menu. A spreadsheet is often inserted to the left of the chosen Worksheet in Excel.

How to Delete a Worksheet

(Right-clicking on the worksheet tab) and choose Uninstall from the menu bar to delete it.

Moving Worksheets (Spreadsheets)

Our worksheets must be in a particular sequence or perhaps in a separate workbook at times.

Moving a Worksheet in the Same Workbook

In the same workbook, there are two methods to transfer a worksheet. The simplest method is to press and (hold) the mouse cursor on the tab of a worksheet and move it to the appropriate place. Keep an eye out for the small black arrows that arise only above. Release the cursor (as it is to the left or right) of the corresponding Worksheet, and the Worksheet would be shifted. Here's an alternative to moving with the cursor if you don't want it. (Right-click the source worksheet's tab) and select ("Move or Copy")... Select the title of the Worksheet you would like the sheet to be placed in the Copy or move browser, then press OK.

How to on Moving a Worksheet to a NEW Workbook

To copy or shift a spreadsheet to a fresh workbook, right-click on the tab and choose "Move or Copy." Tap the drop-down arrow below "To Book:" in the Move or Copy browser, then select New Book. Excel deletes the Worksheet from the current workbook and generates a fresh one from the modified Worksheet. Also, save the workbook to your device.

"How to" Move a Worksheet to a Different Workbook

Both the target and source workbooks should be accessible. Right-click on the source worksheet's tab where you want to pass and choose "Copy or Move." Then, at the top of the page, under "To Book," press the small down arrow to open the drop-down menu, and choose the desired workbook's name where the Worksheet is to be shifted to. Check to see whether the Worksheet was correctly transferred to another workbook, then save the workbook.

Copying Worksheets (Spreadsheets)

It's always simpler to copy and then change a current worksheet instead of starting fresh, especially if you'll just use the same several formats (formulas), etc.

Copying a Worksheet in the Same Workbook

Do it by Right-clicking on the tab of the Worksheet and choosing "Move or Copy." to copy the Worksheet into the same workbook. Check the "make a copy" box in the Move or Copy browser, then press the title of the Worksheet where you'd like the sheet to be placed before clicking OK.

Copying a Worksheet to a NEW Workbook

Right-click on the tab of the source worksheet and choose "Move or Copy." to copy the worksheet into a new workbook. Press the fall arrows below "To Book:" in the Move or Copy browser, then select New Book. Excel created a fresh workbook with the copy (spreadsheet) in it and saved the latest workbook to your device.

Copying a Worksheet to a Different Workbook

Below is the simplest method to (copy a worksheet) to some other workbook. Both the target and source workbooks should be accessible. Right-click on the source worksheet's tab (the one you want to copy) and choose the "Move or Copy," "Make a copy" box, and check at the bottom of the Copy or Move browser. At the top of the screen, below "To book," press the small down arrow to activate the drop-down screen and choose the target workbook's names (the other workbook). Check to see whether the Worksheet was correctly copied to some other workbook before saving it.

You should paste and copy the details as follows with a less sticky solution. (Right-click) in the upper left side cell in the (source worksheet) to pick all of the workbook's cells and press Copy.

Then, in the other Excel workbook, identify a blank worksheet, right-click in the upper left side cell to pick all cells, and (paste). Save the workbook to your device.

To delete the animation border from the first (source) Worksheet, click the ESC key and press in an empty cell to deactivate all of the cells.

What are Command Groups

Various command groups will display below based on which tab you press; house, insert, page style, etc.

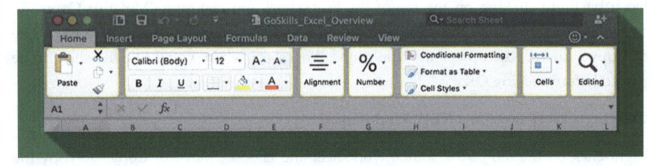

What Activities Can You Do with Excel

Excel is a strong and versatile tool that helps you to organize, analyze, present, and even automate your files.

Spreadsheets are typically used in three steps:

- Enter Data.

- Do something with data.

- Interpret that data.

- And, on rare occasions, there is a fourth step:

- Automating the process

These steps are explained below.

Enter Data

Simply enter data in an excel sheet what you want to do.

Do Something with the Data

Try styling the spreadsheet before doing something about it to make it easy to use. Here are a few Excel functions you may be familiar with.

Conditional Formatting

Although formatting is important for easier reading, it is not the only advantage.

Have you ever come across the term "conditional formatting"? Conditional formatting takes it a step further by styling cells differently depending on what's in them.

For instance, in the previous case, we have a sheet with a set of distilleries and their founding dates. If you want to visit the oldest distilleries, you will use conditional formatting to highlight the three oldest in your spreadsheet.

Note: If the conditional formatting command isn't visible in your Ribbon, look in the upper left-hand corner of your keyboard. Select "More Commands" from the downward arrow button. In the scrolly folder, find the "Conditional Formatting" choice and press the arrow to connect it to your Quick Access Toolbar or Ribbon. The order should then be used from there.

Here's how you would use conditional formatting to draw attention to the oldest distilleries.

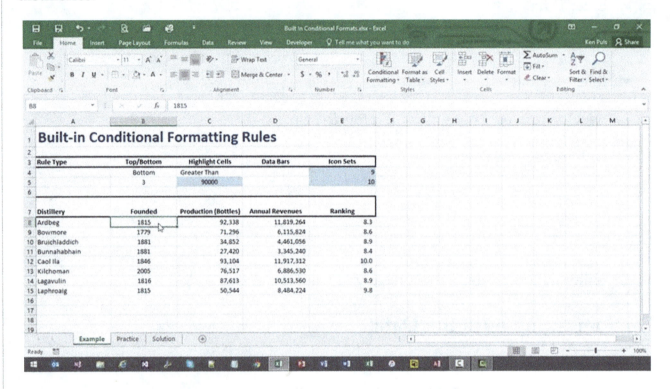

If Excel doesn't already have the choice you want, you may create your own.

It's crucial to remember that the style you prefer would be added to all of the cells you pick, so pay attention to which ones you choose because it will influence how the style looks.

Freeze Panes

This was the subject of the latest analysis. Here's an example:

Nothing is more annoying than scrolling through a large spreadsheet and having to constantly move to the top to see what the column headers are.

Actually, you can allow the column headers and row numbers to remain put because you can notice them no matter where you go in the spreadsheet. Using Excel's helpful "freeze panes" tool, you can accomplish this.

Here's how it's done:

- Select the row below your column headers by clicking on it.

- Select the "View" button.

- Toggle the "Freeze Panes" switch on.

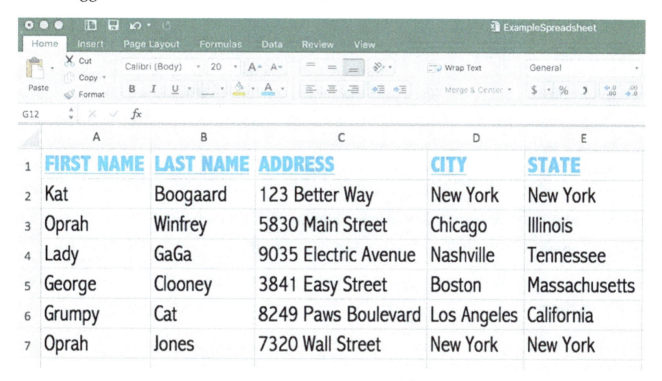

You will see that the details you need are still visible when you scroll down and through your spreadsheet.

Functions and Formulas

Excel's ability to perform advanced calculations and develop business intelligence applications is one of the reasons it is so commonly utilized in the business community.

To do so, you will need to be familiar with formulas and functions. In Excel, formulas are always the most common way to do math.

- ⬚ = ⬚ Each formula begins with = sign

- ⬚ + ⬚ If you want to add, you will use + sign

- **-** If you want to subtraction, you will use - sign

- ***** If you want to multiply, you will use the * sign

- **/** If you want to divide, you will use / sign

You might do as I did above and enter the same numbers, but that's not the best option. What is the reason for this? Since the formula is hard-coded with numbers, you won't be able to copy and paste it into the remaining cells.

But what if you try to do more complex math? How can you go about doing it?

You would use functions, for example. Functions are predefined formulas that are classified according to their functionality—here's a list of some of the most common examples in this case.

SUM Function

Values are added. Specific values, cell references, and ranges, as well as a combination of the three, can be added.

Average Function

The average of the arguments is returned (arithmetic mean). If the set A1:A20 includes numbers, formula = AVG (A1: A20) that returns the average of the numbers in that range.

MAX Function

The highest value in a series of values is returned.

MIN Function

The smallest number in a series of values is returned.

COUNT

The number of cells that contain numbers is counted.

Interpret Your Data

Once you have collected your data, you will need to interpret it.

The problem is that making sense of disorganized data is complicated. As a result, you will have to visualize it first. Here are some suggestions for improving the appearance of your results.

Sort Data by Column

This is how you can filter the data by column/category and make it more presentable.

That's what you should get. Here's how to do it.

Prepare your Data for Sorting

You have to ensure that the data is able to be sorted before you can organize it. If the following statements are valid, your data is ready to be sorted:

Your Data Is Shown in a Tabular Style

There is no empty space between the data you want to filter and the data you would like to sort.

Your data is separated by a header row. (This is optional, but it is strongly recommended.)

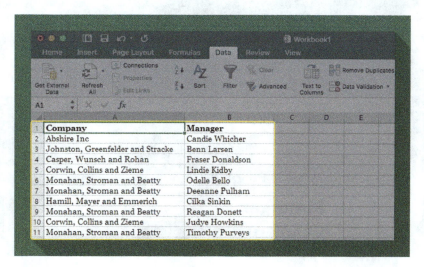

How to Sort on a Single Level

- Every cell in your data table may be selected (or the whole range to be sorted).

- Select data> Sort from the drop-down menu.

- Configure how you want it to be sorted, then press OK.

How to Sort at Multiple Levels

Multi-level kinds are useful with massive data sets because you'll almost always want to filter the data by one column before moving on to the next.

Every cell in your data table may be selected (or the entire range you want to sort).

- Select data> Sort from the drop-down menu.

- Con[the primary type to look the way you like it to.

- Select Add Level from the drop-down menu.

- Configure the secondary type to sort the links the way you like.

- You will keep adding sorting stages as long as you like.

How to Use Filtering

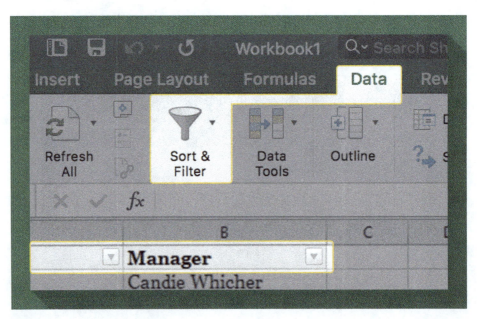

Filtering will help you drill down into details and identify the documents that are relevant to the case you're looking into.

To filter out duplicate data, you must first prepare the data for filtering make sure of the following:

- Your data is shown in a tabular style.

- There is no empty space between the data you want to filter and the data you would like to sort.

- Your data is separated by a header row. (This is optional, but it is strongly recommended.)

- Choose the first row of your files. Select Filter from the Data page (Your headers will now have a row of drop-down arrows).

Filter Specific Words

- Search for specific words using a filter.

- To sort a list, click the drop-down arrow next to it.

- Type the word you're searching for in the Search window, then press OK.

Filter Specific Dates

- Use the date filter to find specific dates.

- To sort a list, click the drop-down arrow next to it.

- To clear unnecessary dates, uncheck the Select everything tag, then use the checkboxes to drill deeply to the documents you want to see.

Filter Multiple Columns

- You may apply filters to several columns to drill down to a smaller number of documents.

Filters May Be Cleared in One of Three Forms

- Press the "Clear filter from" option in the toolbar after clicking the filter icons on the column headers.

- Click the Clear icon, on the Data tab, in the Sort & Filter group.

- Switch off filtering by pressing the Filter icon on the Data tab, then re-add the Filter.

How to Automate Data in MS Excel

If you have repetitive activities to complete, you can use Macros & Scripts to simplify the operation.

What Is the Concept of Macro

A macro is a compilation of pre-recorded acts that you can replicate as many times as you like to avoid having to repeat the same steps manually. In Excel, you can capture a macro for some of the most popular activities and then repeat them to save time.

Macros to Record Macros

- The Developer tab should be added to your Ribbon (if you don't have that on your Ribbon)

- Choose "Customize Ribbon" from the context menu when you right-click every command on the Ribbon.

- In the right-hand list, check the box next to the Developer tab.

- Return to Excel and enable the Developer tab by clicking OK.

Prepare to Record

Since the macro recorder will monitor your errors, practice the measures you'll be doing. Keep in mind that the Macro will perform the same action anytime it is run (This can be changed; however, it would include programming skills).

Record a Macro

- Select "Record Macro" on the developer page.

- Give the Macro a label (no spaces) and, if needed, a capital letter (like R) in the Shortcut box.

- Select This Workbook for the Macro.

- Perform the functions that the Macro is supposed to perform.

- Select "Stop Recording" on the Developer page.

Run the Macro

You have three options for running the Macro:

- To use the keyboard shortcut, press Ctrl + shift + R (or any letter you want).

- To execute the Macro from the macro dialogue, press Alt+F8.

- Insert a Form Control Button on the Worksheet by going to the Developer tab and selecting it (it will prompt you to connect a macro to it). The Macro can then be run by clicking the button.

FUNCTIONS AND FORMULAS

Basic Formulas and Functions

For beginners to become extremely skilled in financial analysis, they must first master the basic Excel formulae. Microsoft Excel is widely regarded as the industry standard in data analysis software. Microsoft's spreadsheet tool is one of the most popular among investment bankers and financial analysts in terms of data processing, financial modeling, and presentation. This will provide you with a rundown of basic Excel functions as well as a list of them.

1. **Include a header and a footer:** We may retain the header and footer in our spreadsheet document using MS Excel.

2. **Replace and Find Command:** MS Excel enables us to locate required data (text and numbers) within a worksheet and replace old data with new data.

3. **Password Security:** Allows users to encrypt their workbooks with a password to protect them from unauthorized access.

4. **Filtering of Data:** Filtering is a quick and simple method of locating and manipulating a subset of data in a range; only the rows that satisfy the criteria you set for a column appear in a filtered range. For filtering ranges in MS Excel, there are two commands:

5. **AutoFilter:** This provides a selection-based filter for simple criteria.

6. **Advanced Filter:** For criteria that are more difficult to define.

7. **Data Sorting:** The process of organizing data in a logical order is known as data sorting. We may sort data in ascending or descending order in MS Excel.

8. **Built-in formulae:** MS Excel has got many built-in formulae for sum, average, minimum, etc. We can utilize such formulas according to our needs.

9. **Create different charts (Pivot Table Report):** MS Excel allows us to make various charts, including bar graphs, pie charts, line graphs, and more. This allows us to evaluate and compare data quickly.

10. **Edits the result automatically:** If any modifications are made in any of the cells, MS Excel immediately updates the result.

11. **Formula Auditing:** We may use blue arrows to graphically illustrate or trace the links between cells and formulae with formula auditing. We can follow the antecedents (cells that supply data to a certain cell) or the dependents (cells that are reliant on a specific cell) (the cells that depend on the value in a specific cell).

There are five popular ways to enter basic Excel formulae while evaluating data. Each method has its own set of benefits. Before getting to the major formulae, let's go through those techniques so you can set up your preferred workflow right away.

Inserting a Formula into a Cell Is a Simple Process

Inserting basic Excel formulae is as simple as typing a formula in a cell or using the formula bar. Typically, the procedure begins with an equal sign followed by the name of an Excel function.

Excel is clever in that it displays a pop-up function suggestion as you start typing the function name. You'll choose your preference from this list. Do not, however, hit the Enter key. Instead, hit the Tab key to continue inserting other selections. Otherwise, you could get an incorrect name error, which looks like '#NAME?'. Pick the cell again and complete your function in the formula bar.

Using the Formulas Tab's Insert Function Option

The Excel Insert Function dialogue box is all you need to complete control over your function insertion. Select Insert Function from the first drop-down option on the

Formulas tab. The dialogue box will include all of the elements you'll need to do your financial analysis.

Formula Tab: Selecting a Formula from One of the Groups

This choice is for individuals who want to get to their favorite features rapidly. To find this option, go to the Formulas tab and select your chosen group. To open a sub-menu with a list of functions, click. You can then choose your preferred option. If your desired group isn't shown on the tab, click the More Functions option, it's most likely buried there.

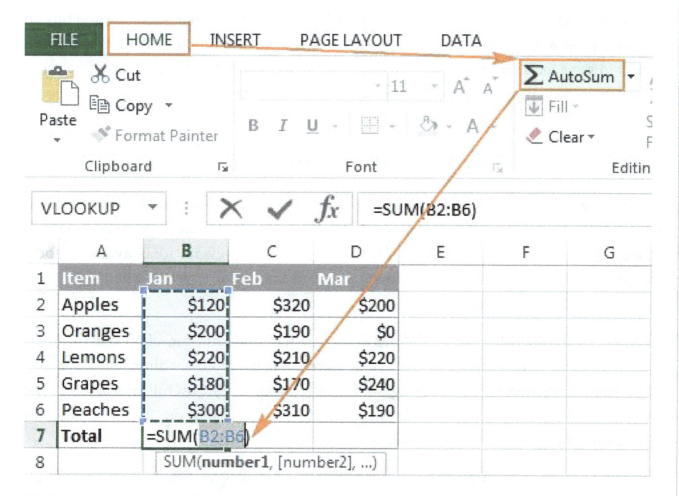

Using AutoSum Option

The AutoSum function is your go-to solution for short and everyday jobs. So, go to the Home page and choose the AutoSum option in the far-right corner. Then click the caret to reveal more formulae that were previously concealed. This option is also accessible after the Insert Function option in the Formulas tab.

Use Recently Used Tabs for a Quick Insert

Use the Recently Used option instead of retyping your most recent formula if you find it tedious. It's on the Formulas tab, right next to AutoSum, as the third menu choice.

You've made the decision. You switched from Evernote to Notion after hearing about the buzzy productivity software. You've learned Notion's fundamentals, built a GTD to-do list, and even a Personal Wiki. As you sit down to create your first If Statement, you expect to see a familiar Excel-like structure, but then it occurs to you; where are the cells? Why aren't these property types all the same? And why do these formulas appear to be so dissimilar?

Advanced Formulas, Features, and Functions

Today's business requires the use of a variety of software and tools to manage and operate effectively. In business jargon, the technologies are utilized to save time and money by providing rapid analytical findings. There are several tools available for a range of companies, the most prominent and well-known of which is Microsoft Office Excel, a tool that every business needs.

Excel's outstanding collection of abilities to execute functions and complex excel formulas is not a myth; it is a declared truth that Excel assists small, medium, and big size organizations in storing, maintaining, and analyzing data into useful information. The program handles and covers different aspects of a business on its own, such as accounting, budgeting, and sales, among others.

Excel is one of the most significant and helpful pieces of software, and its prominent features allow you to perform the following.

Prepare Magnificent Charts

An excel sheet, as we all know, has a large number of grids. These sheets are restricted to numbers or data input, but they may be used to visualize prospective data using advanced Excel formulas and functions. The data is sorted, filtered, and structured using the functions assigned to the rows and columns. As a visual display for greater understanding, the information from the assigning and arranging is transformed into charts.

The set of statistics is difficult to comprehend and draw a conclusion from. The pie charts, grouped columns, and graphs make analyzing and interpreting data in a short amount of time a lot simpler. Excel is a powerful tool for producing company reports and marketing materials.

Conditional formatting is made easier with visual aids. Colors, tints, italics, bold, and other formatting choices help distinguish rows and columns so you can find data quickly and save time. Because of the color difference, a user can quickly identify the appropriate column and row in a large data set. The formatting tab makes it simple to enter a color scheme.

Identifying Trends

When it comes to developing a plan by observing patterns and forecasting the next, the statistical effect in a firm is enormous. Average lines can be given to charts, graphs, clustered columns, and other visual representations. The average line aids an analyst in determining the main trend in the data set. It can quickly comprehend the format's main points.

The projection may be used to take the trend or average lines a step farther. Future trends can be predicted using these projections in the visual depiction. The prediction can aid in the creation of new initiatives that will propel the company to new heights of success.

Bring Anything

The software's versatility may handle almost any form of data. Spreadsheets, documents, and even pictures can be used. Access is made easier when all of the data is placed under one roof for convenience. In Excel, importing any data is a breeze. The Insert Tab assists the user in data aggregation.

Excel's cloud function has elevated its use to new heights. Office 365 Business and its premium edition may be accessed from various devices, making it easier to do business. This software allows for remote working by coordinating papers and sheets.

Advanced Excel Formula and Functions

Excel offers a plethora of useful uses. The simple form is used by 95% of the users. For sophisticated computations, there are functions and advanced excel formulas that may be employed. The functions are meant to make it simple to search up and prepare a vast amount of data, whereas the advanced excel formula is used to extract new information from a specific data collection.

Excel's OFFSET Formula

This sophisticated excel function, when used with other functions like SUM or AVERAGE, may give computations a dynamic feel. It's excellent for putting continuous rows into a database that already exists. OFFSET Excel provides us with a range to fill in with the reference cell, the number of rows, and the number of columns. Ex. If we have a worker's salaries list sorted by employee ID and want to calculate the average of the top five employees in the company, we may use the formula below. The formula below returns the salary results every time.

Excel formulas for LEFT, MID, and RIGHT

This sophisticated excel formula may be used to extract a specified substring from a text. It could be appropriate for our needs. Ex. We may apply the LEFT formula in Excel with the column name and second parameter as 5 to extract the first five characters of the employee's name.

Excel's CONCATENATE Formula

One of the equations that may be utilized with numerous versions is this excel advanced function. This sophisticated Excel formula allows us to combine many text strings into a single one. For example, suppose we wish to display both the employee ID and the employee name in a single column.

SUMIF Formula in Excel

When using the sum or count function in some studies, you may need to filter some observations. This sophisticated excel SUMIF function in excel comes to our rescue in such situations. After filtering them based on the conditions specified in this sophisticated excel formula, it adds up all of the observations. What if we want to know the total salary of workers with employee IDs higher than 3?

IF AND Formula in Excel

There are numerous situations where flags must be created depending on limitations. The basic syntax of IF is familiar to all of us. This advanced excel IF function makes a new field based on a constraint on an existing field. But what if we need to take into account many columns while constructing a flag? Ex. In the example below, we want to identify all workers with a salary of more than $50,000 but an employee ID of more than three.

Excel MATCH Formula

When there is a certain text or number in the provided range, this Excel advanced formula provides the row or column number. In the example below, we're looking for "Rajesh Ved" in the Employee Name field.

Excel VLOOKUP Formula

One of the most often used formulae in Excel is the advanced excel function. It is mostly due to the formula's simplicity and its use in seeking a certain value from other tables that share a similar variable. Assume you have two tables with pay and name information for a company's employees, with Employee ID as the main column. In Table A, you wish to extract the salary from Table B.

VLOOKUP is divided into three categories:

1. You cannot have the main column to the right of the column for which you want the value from another table to be populated. Employee Salary cannot be placed before Employee ID in this scenario.

2. The first value in the duplicated values in Table B's main column will be filled in the cell.

3. If you add a new column to the database (for example, before Employee Salary in Table B), the formula's output may change depending on the position you specified in the formula.

Excel INDEX Formula

This sophisticated Excel formula is used to determine the value of a cell in a table depending on the number of rows, columns, or both in the table.

CHARTS AND GRAPHS

What Is a Chart

I t is a representation of data in both rows and columns in a visual format. Charts are often used to analyze data sets for patterns and trends. Suppose you have been keeping track of sales data in Excel for the last three years. You will easily see which year had the least sales and which year had the most when looking at charts. You may also use charts to compare defined goals to actual accomplishments.

Graphs and charts enable a better understanding of the results by visualizing numeric values in an easy-to-understand way. Even if words are being frequently interchanged, they are distinct. Graphs are the simplest graphic representation of data, and they usually show data point values over time. Charts are more complicated since they help you to compare parts of a data set to other data of the same set. Charts are often more attractive than graphs because they often have a different shape from a standard x- and y-axis.

In presentations, graphs and charts are often used to provide a brief snapshot of clients, development, or outcomes to management or team members. You can make a graph or chart to display almost every kind of quantitative data, saving you the effort and time of searching through spreadsheets to identify relationships and patterns.

Excel makes it simple to create graphs and charts, particularly when you can store the data in an Excel Workbook rather than importing it from another application. Excel also comes with several pre-made graphs and chart types from which you can choose the one that better reflects the data relationship you want to emphasize.

When to Use Each Chart and Graph Type in Excel

Excel has a huge graph and chart library to help you visually display your results. While several chart forms may function with a given data set, it is essential to choose what better suits the message you want to tell with the data. You should, of course, apply graphical elements to a graph or chart to enhance and modify it.

Column Charts

Column charts are ideal for comparing data or several categories with a single variable (for instance, multiple genres or products). There are five major types of charts or graphs in Excel.

1. Clustered

2. Stacked

3. 100% stacked

4. 3-D clustered

5. 3-D stacked

6. 3-D 100 % stacked

7. 3-D

These are the seven-column chart forms available in excel.

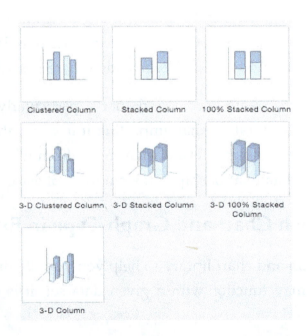

Choose the visualization that best tells the message of your results.

Bar Charts

The major difference between a column chart and a bar chart is that the bars in a bar chart are horizontal instead of vertical. While column charts and bar charts may also be used interchangeably, some people choose column charts when dealing with negative

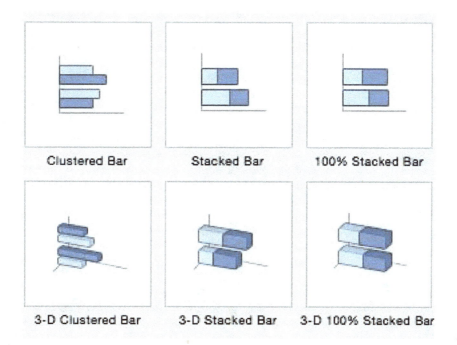

Clustered Bar   Stacked Bar   100% Stacked Bar

3-D Clustered Bar   3-D Stacked Bar   3-D 100% Stacked Bar

values because it is simpler to represent negatives vertically on a y-axis.

Pie Charts

To contrast percentages of a whole (the sum of your data values), use pie charts. Each value is expressed by a pie slice, allowing you to see the proportions. Following are the five types of pie charts:

- Pie

- Pie of pie (which divides one pie into two to indicate sub-category proportions)

- A bar of pie

- 3-D Pie

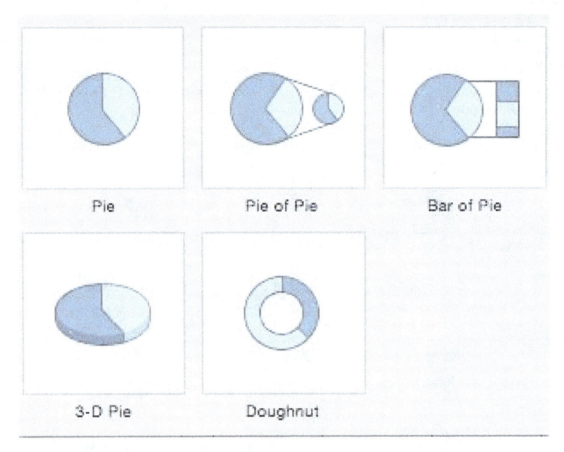

- A doughnut.

Line Charts

Instead of static data points, a line chart is best for displaying trends over time. The lines link each data point, allowing you to see if the value(s) decreased or increased over time.

- Line

- Stacked line.

- 100% Stacked line

- Line with markers

- Stacked line with markers.

- 100% Stacked line with markers

- 3-D line

These are the seven-line chart choices.

Line	Stacked Line	100% Stacked Line
Line with Markers	Stacked Line with Markers	100% Stacked Line with Markers
3-D Line		

Scatter Charts

Scatter charts are used to display how one variable influences another. They are similar to line graphs in that they help display changes in variables over time. (This is referred to as correlation.) Bubble charts, which are a common chart type, are classified as scatter.

- Scatter

- Scatter with smooth lines and markers

- Scatter with smooth lines

- Scatter with straight lines and markers

- Scatter with straight lines

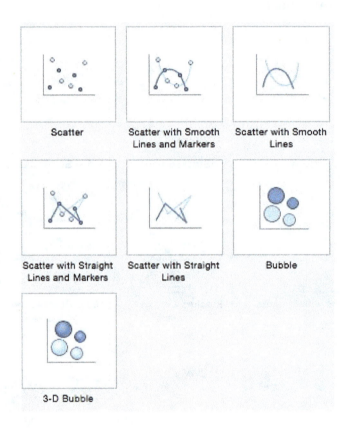

- Bubble

- 3-D Bubble

These are the seven scatter chart choices.

In addition, there are four minor types. These are more case-specific in use:

Area

Area charts, like line charts, represent shifts in values over time. On the other hand, area charts are useful for highlighting differences in change between multiple variables since the area under each line is solid.

- Area

- Stacked area

- 100% Stacked area

- 3-D Area

- 3-D Stacked area

- 3-D 100% Stacked area.

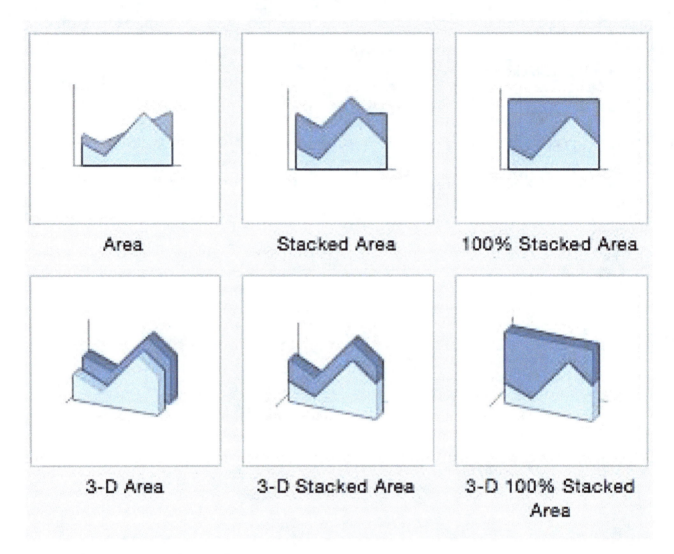

Area   Stacked Area   100% Stacked Area

3-D Area   3-D Stacked Area   3-D 100% Stacked Area

These are the six types of area charts.

Stock

Investors use this form of chart to show the low, high, and closing price of a stock and financial analysis. However, if you choose to represent the range of a value (or the range of its expected value) and its exact value, you can use them in every case. Choose from stock chart options such as:

- High-low-close

- Open-high-low-close

- Volume-high-low-close

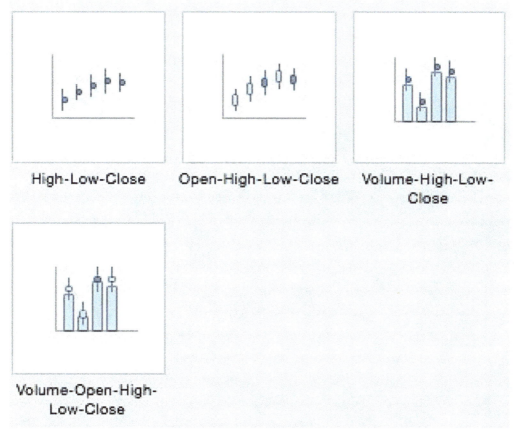

High-Low-Close   Open-High-Low-Close   Volume-High-Low-Close

Volume-Open-High-Low-Close

- Volume-open-high-low-close.

Surface

To represent data over a 3-D landscape, use a surface chart. Big data sets, data sets of more than two variables, and data sets with groups inside a single variable benefit from the additional plane. Surface charts can be difficult to understand, so make sure the audience is comfortable with them.

- 3-D Surface

- Wireframe 3-D surface

- Contour

- Wireframe contour

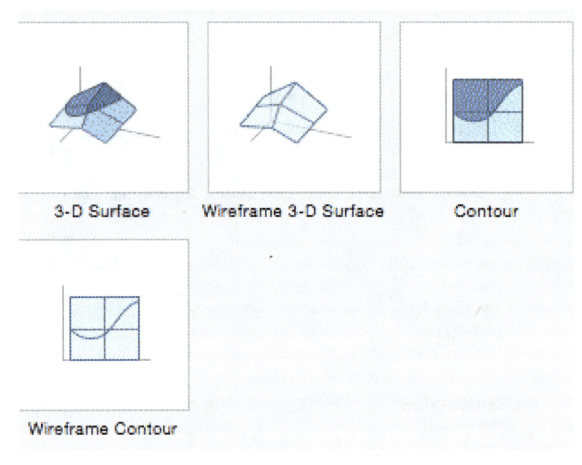

3-D Surface   Wireframe 3-D Surface   Contour

Wireframe Contour

These are its types.

Radar

A radar chart is useful for displaying data from different variables in relation to one another. The central point is the starting point for all variables. The trick to using radar charts is to compare all particular factors in relation to one another; they are often used to compare the weaknesses and strengths of various products or employees.

- Radar

- Radar with markers

- Filled radar.

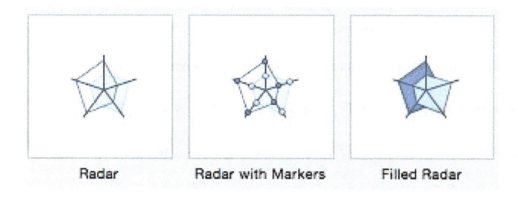

Radar   Radar with Markers   Filled Radar

These are the three kinds of radar charts.

Waterfall

A waterfall chart, which is a set of column graphs that display negative and positive changes over time, is another popular chart. A waterfall chart does not have an Excel preset, so you can download a guide to make the task simpler.

The Significance of Charts

- A chart allows the user to see the data as a visual representation.

- It is easy to understand compared to data in cells.

- It is easier to analyze patterns and trends in the charts.

Step by Step Example of Creating Charts in Excel

This tutorial will build a basic column chart that shows the sold quantities versus the sales year.

- Open your Excel.

- Enter the data.

Your workbook would look like this:

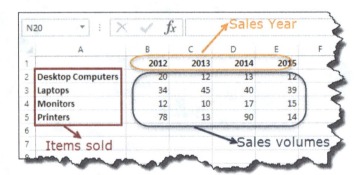

To get the chart of your desire, follow these steps.

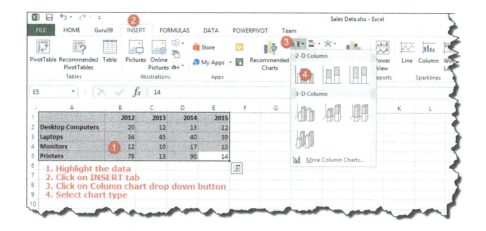

- Select all the data that you want to show in your graph.

- From the ribbon, click the INSERT tab.

- Click on the "Column chart" drop-down button.

- Click on the chart type that you want.

You will see a chart like this:

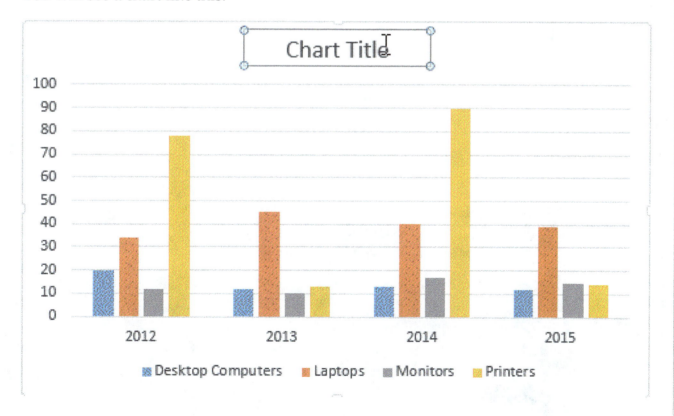

When you select your desired chart, the ribbon will activate the following tab.

Try to use different chart types and styles and other options offered in the chart.

PIVOT TABLE

Creating Pivot Table

Create an Excel table to arrange the data into rows and columns before creating a pivot table. You may build a pivot table once you've prepared your source data. First, check which pivot table designs Excel recommends.

Any cell in the source data table may be selected.

- Select the Insert tab from the Ribbon.

- Click Recommended PivotTables in the Tables group.

- Scroll down the list in the Recommended PivotTables box to view the recommended layouts. To see a bigger version of a layout, click on it.

- Suggested pivot tables

- Click OK after selecting the layout you wish to use.

Filtering

AutoFilter is the fastest and most convenient method to filter data in a table (or list). When we say "filter," we're simply saying that we're temporarily concealing the rows we don't want to view. This feature shows drop-down lists at the top of each column and enables the user to filter the data by selecting unique values within each column. These drop-down lists may then be used to filter the data in your table, either singly or in combination. When you "filter" a table, any rows that don't match the value you selected from the drop-down list are buried.

The filters available in each column are determined by the data category. Number Filters, for example, are available in columns with numbers, whereas Text Filters are available in columns with text, and Date Filters are available in columns with dates. In each category, there are several built-in sorting features that may be useful:

Date Filters

These filters are very durable. Data may be sorted by day, month, year, week, and a quarter.

Text Filters

The most useful default text filters are Begins With, Contains, and Does Not Contain. You may build a new filter if none of the current filters meet your requirements. We'll accomplish it immediately with the help of Dates.

Now, click the arrow next to the column you wish to sort, then choose Custom Filter from the Filters for that column. The Custom AutoFilter box that appears is extremely user-friendly. Make a phrase that explains the filtering you want using the drop-down options.

Number Filters

Top 10, Below Average, x, and Above Average are the most helpful for businesses.

Change Summary Calculation

When you build a pivot table report, Excel defaults to collecting or summing the elements to summarize the data. You may wish to use functions like Min, Max, and Count Numeric instead of Sum or Count. There are a total of 11 choices available. However, the most frequent cause for changing a summary computation is that Excel has opted to count rather than total your data erroneously. When you insert a numerical field to the Values section of a pivot table, the default summary function is Sum or Count. The default function cannot be altered; it is determined by the values of the field:

- The sum will be the default if the field includes numbers.

- The count will be the default whether the field includes text or blank cells.

To choose a different summary function once a field has been added to the pivot table, follow these steps:

- Change a cell in the Value field by right-clicking on it.

- Select Summarize Values from the pop-up menu. By

- Select the Summary Function you wish to use by clicking on it.

Sorting Data by Specific Trait

Whenever you are sorting data, it is critical to determine whether you want the sort to apply to the whole worksheet or just a certain cell range.

Sort Range

In a range of cells, you may sort the information in that range, which can be useful when dealing with a sheet that includes a number of tables. The sorting of a range has no effect on the rest of the worksheet's information. It is possible that the default sorting choices will not be able to arrange data in the order that you need it to be. Excel, on the other hand, enables you to build a custom list in which you may choose your own sorting order.

- Choose a cell in the column you wish to sort by and press Enter.

- Select the Data tab and then the Sort command to sort the data.

- On the Data tab, there is a Sort button.

- The Sort dialogue box will be shown after that. Then, in the Order box, choose Custom List from the drop-down menu for the column you wish to sort by.

- The Custom Lists dialogue box will be shown after that. Choose NEW LIST from the Custom Lists drop-down menu.

- Fill in the List entries box with the items in the custom order you want them to appear.

Sort Sheet

One column is used to arrange all of the information in your worksheet. When the sort is performed, all of the information that is related across each row is retained together.

Fine Tune the Calculator

You already know that when you enter a field to the Values box, Excel makes a guess as to what calculation you want to conduct on the data. Generally, it is assumed that you wish to execute a sum operation on the field, which totals all of the values in the field. This computation, on the other hand, is not necessarily the correct one. Because Excel makes it simple to alter the kind of calculation you're doing, you can save time. In fact, as you'll see in the next sections, you may use the same pivot table to conduct several calculations at the same time, as well as to include custom formulae into the mix. You may change the computation that the pivot table performs by following the steps below:

- Locate the relevant field in the Values box of the PivotTable Field List pane by typing the field name into the search box. By selecting Value Field Settings from the drop-down arrow, you may customize your fields.

- The "Sum of Quantity" item in the Values box may be changed if you wish to alter the current operation, which is the summation of the Quantity values for each row in a group.

- Select a different choice from the drop-down menu under the "Summarize by" header.

- If you wish to specify an alternative number format for the summary information to be shown, click the Number Format button, choose a new format, and then click the OK button.

- When you choose Number Format, Excel displays a condensed version of the Format Cells dialogue box, which has just the Number tab and no other tabs.

- The number of decimal places may be changed here, as well as the currency symbol, among other things, and so on.

- To dismiss the Value Field Settings dialogue box, press the OK button.

- As soon as new information is entered into Excel, the pivot table is updated.

Macros

Macros are pieces of code that automate work in a program, which saves and records your work and accomplish what you exactly need to do, quickly and with a single click of a button, without having to learn a new language. Macros are a type of code that is used to automate work in a program. When working with a spreadsheet program such as Excel, macros may be very useful. They are more powerful than the usual functions you put into a cell (for example, =IF(A2100,100, A2), which are hidden behind the regular user interface. These macros make Excel do the heavy lifting for you. They take the role of operations that you would normally do manually, such as formatting cells, copying data, or computing totals.

As a result, you may rapidly replace monotonous activities with a few clicks. Create macros by simply recording your actions in Excel and saving them as repeated steps, or you may use Visual Basic for Applications (VBA), a basic programming language that is integrated into Microsoft Office, to create macros that are more complex. Understanding how to automate Excel is one of the simplest methods to make your job more efficient, particularly given the fact that Excel is utilized in so many different work processes.

Assume that you export analytics data from your content management system (CMS) once a week in order to produce a report on your website. The only issue is that those data outputs aren't always in an Excel-friendly format, which may be frustrating. They're jumbled, and they often include much more information than is necessary for your report. To do so, you must clear up empty rows, copy/paste data into the appropriate locations, and build your own charts to display data and make it print-friendly.

It may take you many hours to accomplish all of these tasks. If only there was a way to click a single button and have Excel handle everything for you in an instant. Is it possible for you to predict what we're going to say next? All it takes is a few minutes to set up a macro, and then that code may be used to do the necessary tasks on a consistent basis. It's not even close to being as tough as it seems.

Importance of the Pivot Table

A pivot table takes a data field that has been provided by the user and turns the head of each column into a data option that can be readily manipulated by the user in the table. Columns holding data may be readily deleted from, added to, or changed about in a table with relative simplicity. Long spreadsheets of raw data may be transformed into user-friendly and useful summaries with this software. The information may be summarized in a variety of ways, including frequencies and averages. There are many advantages, which are detailed here, to using a pivot table in Excel.

Easy to Use

The fact that pivot tables are simple to use is a significant benefit. By moving columns to various parts of the table, you may quickly summarize data. With a click of the mouse, you may rearrange the columns in any way you like.

Easy Summary of Data

Another major advantage of pivot tables is that they make it simple to summarize data out of thousands of rows and columns of unstructured data, the table aids in the creation of a succinct summary. You can condense a lot of information by using these tables. The information may be summarized in an easy-to-understand manner. Users may name and organize the data in whatever manner they choose, and they can rearrange the rows and columns to suit their requirements.

Easy Data Analysis

Excel pivot tables allow you to manage huge amounts of data in a single operation. These tables enable you to deal with a huge quantity of data while only seeing a few data columns. This makes huge amounts of data easier to analyze.

Find Data Patterns

Pivot tables in Excel enable you to build customized tables from huge data sets. This kind of data manipulation will aid in the discovery of any recurrent patterns in the data. As a result, precise data forecasting will be easier.

Helps in Quick Decision Making

A pivot table is a useful Excel reporting tool because it enables users to evaluate data and make choices based on it quickly. This is a significant benefit in the industrial sector, where making accurate and fast choices is critical.

Quick Report Creation

A few of the most useful aspects of excel pivot tables is that they make it easier to generate reports. This reduces the need for you to spend lengthy and exhausting hours manually generating reports. Aside from that, the table allows you to include connections to external sources in the report you've prepared.

TOOLS TO SAVE TIME

L et us explain how to have some of your time back if Microsoft Excel has taken up a lot of it in the past. These straightforward instructions are essential to remember. Microsoft Excel is a strong, feature-rich workbook and spreadsheet program and uses features that save you time, whether at home or work; it allows you to move on to the next task quicker.

Here are a few fast ways to get through your spreadsheets, workbooks, and other similar activities.

Time Saving Templates

The usage of a template is among the most effective time savers for almost every project. You will use them to create meeting schedules, newsletters, and reports in Microsoft Word, and Templates are almost as effective in Excel. Using these time-saving pre-formatted resources for project calendars, schedules, budgets, invoices, and more will save you a lot of time.

It's just as easy to have the models you need as it is to use them. Excel models for schedules, calendars, timesheets, and cost trackers are available on the Microsoft website. Schedules, forecasts, balance sheets, and to-do lists are also available from Vertex42. Spreadsheet 123 is another excellent resource for inventory, invoices, attendance forms, and sign-in sheets.

Freezing Columns and Rows

If you have a long spreadsheet with many numbers, you might have to navigate down or even through to see anything. This ensures you'll lose track of the headings and have to keep scrolling back to locate them. If you freeze the columns and rows, though, the headers can stay in place while you pass around the spreadsheet.

In the View Tab, you can find the Freeze Panes of the ribbon. Pick Freeze First Column, Freeze Top Row, or both if desired from the Freeze Panes dropdown.

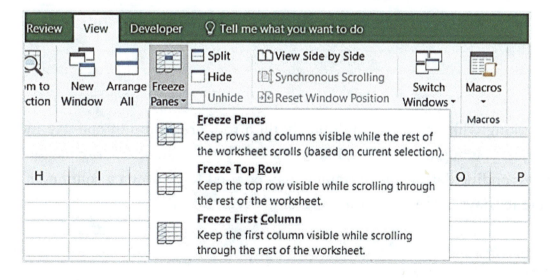

You will find that the headers are already available whether you scroll up, down, left or right. Pick the Freeze Panes command again and press Unfreeze Panes to unfreeze certain columns and rows.

The procedure is somewhat different in older versions of Excel. Click Freeze Panes after selecting the cell that is unique to both the column and row you choose to freeze.

Operating Fill Handle

When you want to populate many cells in Excel, the Fill Handle will save you a lot of time, and there are a few various ways to use it. To begin, enter 1 in the first cell & 2 in the cell below it to quickly construct a list of numbers that add up. After that, pick all cells and drag the fill handle to populate the cells as desired.

Monday		January		1	
Tuesday		February		2	
Wednesday		March			
Thursday		April			
Friday		May			
Saturday		June			

When working with days, you can quickly fill a row or column in one-day increments. For example, you might type 25/12/16 into the cell, pick it, and then drag the fill handle to add corresponding dates. This technique may be used to go down a column or around a board and for days of the week & months of the year.

Monday		1	January		
Tuesday		2			
Wednesday		3			
Thursday		4			
Friday		5			
Saturday		6			

If you need to fill an entire column in your spreadsheet with the same value, the fill handle comes in handy. Pick the cell & double-click the fill handle to insert a word or letter, for example. This will populate the remaining cells of the column with that value before your spreadsheet runs out of data.

Transposing Rows and Columns

If you have a spreadsheet with headings in rows or columns or both and decide they will fit best the other direction, changing them is easy. This eliminates the need to retype such headings. To switch cells from row headings to column headings, follow these moves.

- Select the headings-containing cells in the column.

- Right-click and choose Copy, or go to the Home tab of the ribbon and click the Copy icon.

- Choose the cell in the column and row where the headings should start.

- On the Home tab of the ribbon, right-click and choose Paste Special, or press Paste & then Paste Special.

- Select the Transpose checkbox at the bottom right corner.

- Click the OK button.

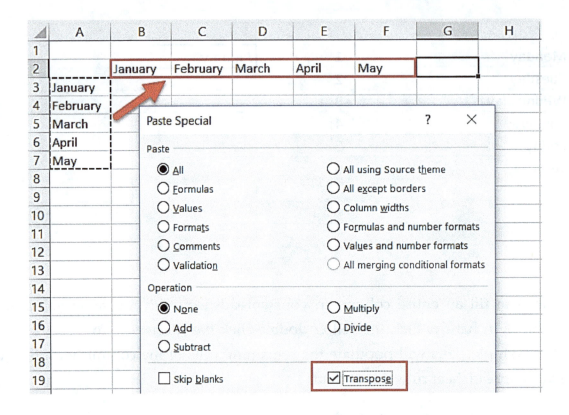

Accessing Calculator

You can use an add-in to add a calculator to the sidebar of the spreadsheet, but you can still use the built-in calculator function. This is a good option when you need to do fast calculations that aren't formulated in your database. You may even connect the calculator to your Quick Access toolbar or your ribbon.

Start by going to File > Options and adding the calculator to any position. Then, based on where you like it, pick Configure Ribbon or Quick Access Toolbar. Select All Commands from the dropdown box in the Select commands from the dropdown box. To add it to the toolbar, scroll down and press Calculator, then touch the Add icon. If you like to put it on your ribbon, you'll have to make a custom group to put it in there.

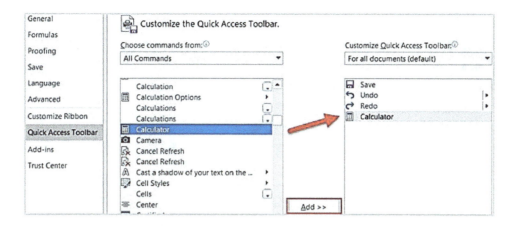

Linking to Cells or Tabs

If your workbook includes several spreadsheets of data that are cross-referenced, having a direct connection allows you to access them quickly. This is particularly useful if you're sharing a workbook and want someone to access the info easily. To make the relation, simply follow these easy steps:

- Choose the cell that contains the data you want to connect.

- Right-click and choose Hyperlink from the context menu, or go to the Insert tab & pick Hyperlink from the ribbon.

- Select Position in this Document in the popup window.

- Then type the text you want to appear in that cell, a particular cell reference if you want and the database in the workbook that contains the details you want to connect to.

- Click the OK button.

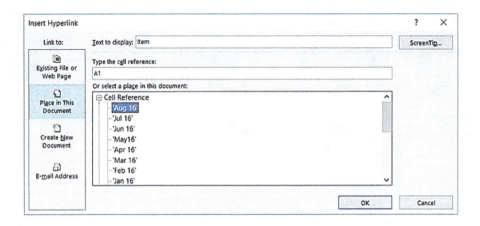

AutoSum

Some people believe that learning how to use formulas in Excel takes so much time. However, also for basic calculations, these built-in functions will significantly speed up the spreadsheet job.

The AutoSum key must be on your Home tab unless you've transferred or removed it from your ribbon. You can access the most basic formulas with a single click. You may incorporate, list, average, or find the minimum or limit of a set of numbers. To choose your formula, simply press the arrow on the AutoSum icon.

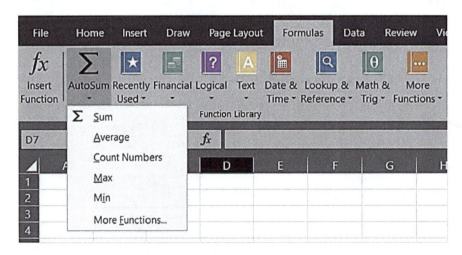

Your Formulas tab includes much more choices than the AutoSum function. Each formula is organized into categories to make it easier to locate. Financial, logical, math, mathematical, and engineering functions are accessible.

The AutoSum function, on the other hand, is simple and easy for the most commonly used formulas.

Conditional Formatting in a Simple Way

Another Excel functionality that many people can find challenging is conditional formatting. It's a fantastic tool for data that you would like to pop off the tab, though.

For example, suppose you have a spreadsheet of survey results and want to see how many Yes responses you have vs. No answers at a glance. These instructions will teach you how to implement basic formatting.

- Select the cells with the Yes/No responses.

- Press the Conditional Formatting dropdown box on the Home page.

- Select Highlight Cells Rules & afterward Text That Contains from the dropdown menu.

- In the left column, type the term Yes, and in the right box, select the formatting for

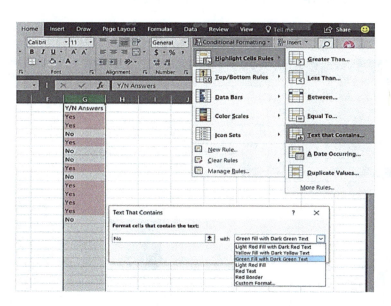

it.

- For the No responses, repeat the procedure.

All of the Yes & No responses will be formatted in the way you specified, making them simple to spot.

If you intend to keep adding results, this conditional formatting may be applied to the whole column or row rather than just a group of cells. In this method, data would be automatically formatted when you access it in the future.

Inserting Charts Quickly

Excel's Charts functionality is a fantastic tool for visually viewing your results. You may also choose from a wide range of chart categories, including line, bar, pie, column, and several more.

With only a few taps, you can attach a map using the conditional formatting for Yes/No responses seen above.

- Select the cells with the Yes/No responses.

- Select Recommended Charts from the Insert page. With this choice, Excel can take your data & place it in the most appropriate chart format.

- If the chart corresponds to you, press OK, and it will be added to your

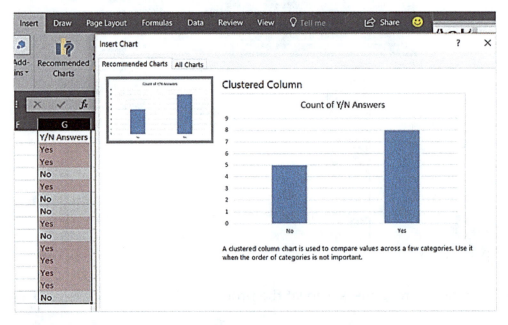

spreadsheet.

The quickest and simplest method for creating an Excel map, and it just takes a minute. If you don't like the chart that was created for you, you can play with other styles by clicking the All-Charts tab in the popup window.

Using Filters for Sorting

When your spreadsheet has a lot of columns of data, you might want to filter or sort it all by one column. Although there are a few options for doing so, using a filter is the fastest and safest method.

- By pressing the triangular button next to the first column on the top left, you may choose the whole sheet.

- Select Sort & Filter from the Home page.

- Select Filter.

That concludes our discussion. Each of the columns would have an arrow in the first row as a result of this quick step. When you press an arrow, you can sort the whole sheet in whatever direction you like through that column. You can order from oldest to newest if it's a date area or alphabetically if it's a text field.

You may even select the details and display just the entries you're interested in. Checkboxes appear next to the entries when you press an arrow. Through checking and unchecking these boxes, you will filter the details to see just what you need.

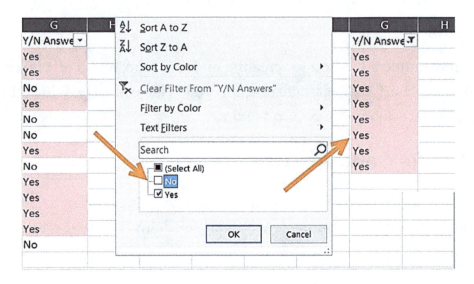

The Filter function has the advantage of having no negative impact on the rest of your results. When you sort a sheet as a whole, Excel adjusts both columns. Excel can simply hide anything you don't want to show if you filter.

Format Painter

You might still be comfortable with the Format Painter if you have used other MS Office applications like Word. With a simple click in Excel, you can add the same formatting through one or even more cells to other cells.

- Choose cell, group of cells, column, or row from which you would like to copy the formatting.

- Select the Format Painter from the Home page.

- Choose cell, group of cells, column, or row to which the formatting should be copied.

- This is a brilliant way to easily extend useful cell formatting to others without having to do some manual work.

Window Switching and Viewing

Is there really a moment when you need to deal with several Excel workbooks at once? You might need to go over results, compare them, or copy them from one workbook to another.

SHORTCUT KEYS

I magine there is a shortcut to getting to a particular place in just 5 minutes, and initially, the path should cost less than an hour. Surely you will choose the shortcuts. If you are using Excel, especially that of 2021, easier, faster, and convenient to use, we will be taking ourselves to learn some tips and tricks which will come in practical when starting to get acquainted with the use of this app.

Excel Shortcuts

Editing Shortcuts:

Keys:	Functions:
• F2	To edit cell
• Ctrl + C	To copy
• Ctrl + V	To paste
• Ctrl + X	To cut
• Ctrl + D	To fill down
• Ctrl + R	To fill right
• Alt+ E+ S	Paste special

- F3   To paste the name into a formula

- F4   Toggle reference

- Alt +Enter   To start another new line within the same old cell

- Shift + F2   To insert or edit a cell comment

- Shift + F10   To display a shortcut menu

- Ctrl + F3   To define the name of a cell

- Ctrl + Shift + A   To insert arguments names with parentheses for a function after typing a function name in a formula

- Alt + I + R   To insert a row

- Alt + I + C   To insert a column

Formatting Shortcuts

Keys: Functions:

- Ctrl + B   To bolden

- Ctrl + I   For italics

- Ctrl + Z   To undo

- Ctrl + Y   To repeat the last action

- Ctrl + A   To select all cells

- Ctrl + 1   To display or bring up the format cell menu

- Ctrl + Shift + !   For number formatting

- Ctrl + Shift + %   For percent format

- Ctrl + Shift + #   For date format

- Alt + h   To increase decimal

- Alt + h+ 9   To decrease decimal

- Alt + h + 6   To increase indent

Navigation Shortcuts

Keys: Functions:

- Arrow   From one cell to the next

- F5   Go to

- Ctrl + Home   Go to cell 1

- Home

  To go to the beginning of a row

- Shift + Arrow

  To select the adjacent cell

- Shift + Spacebar

  To select an entire row

- Ctrl + Spacebar

  To select an entire column

- Ctrl + Shift + Home

  To select all to the start of the sheet

- Ctrl+ Shift + End

  To select all to the last used cell of the sheet

- Ctrl + Shift + Arrow

  To select the end of the last used row/column

- Ctrl + Arrow

  To select the last used cell in rows/columns

- PageUp

  To move the screen up

- PageDown

  To move the screen down

- Alt + PageUp

  To move the screen to the left

- Alt+ PageDown

  To move the screen to the right

- Ctrl + PageUP/Down

  To move the next or previous worksheet

- Ctrl + Tab   To move to the next worksheet while on the spreadsheet

- Tab   To move to the next cell

File Shortcuts

Keys: Functions:

- Ctrl + N   New

- Ctrl + O   To open

- Ctrl + S   To save workbook

- F12   Save As

- Ctrl + P   Print

- Ctrl + F2   To open the preview print window

- Ctrl + Tab   To move to the next workbook

- Ctrl + F4   To close a file

- Alt + F4   To close all open Excel files

Paste Special Shortcuts

Keys: Functions:

- Ctrl + Alt + V+T   Paste Special formats

- Ctrl + Alt + V+V   Paste Special values

- Ctrl + Alt + V+F   Paste Special formulas

- Ctrl + Alt + V+ C   Paste Special comments

Ribbon Shortcuts

Keys: Functions:

- Alt   To show ribbon accelerator keys

- Ctrl + F1   To show or hide the ribbon

Clear Shortcuts

Keys: Functions:

- Delete   To clear cell data

- Alt+ h + e + f   To clear cell format

- Alt+ h + e + m   To clear cell comments

- Alt+ h + e + a   To clear all data formats and comments

Selection Shortcuts

Keys: Functions:

- Shift + Arrow   To select a cell range

- Ctrl + Shift + Arrows   To highlight a contiguous range

- Shift + PageUp   To extend selection up one screen

- Shift + PageDown   To extend selection down one screen

- Alt + Shift + PageUp   To extend selection left one screen

- Alt + Shift + PageDown   To extend selection right one screen

- Ctrl + A   Select or highlight all

Data Editing Shortcut

Keys: Functions:

- Ctrl + D   To fill down from cell above

- Ctrl + R   To fill right from cell left

- Ctrl + F   To find and replace

- F5 + Alt + s +o   To show all constants

- F5 + Alt + s +c   To highlight the cell with comments

Data Editing (Inside a Cell) Shortcuts

Keys: Functions:

- F2   To edit the active cell

- Enter   To confirm a change in a cell before opting out of that cell

- Esc   To cancel a cell entry before opting out of that cell

- Alt + Enter   Insert a line break within a cell

- Shift + Left/Right   To highlight within a cell

- Ctrl + Shift + Left/Right   To highlight contiguous items

- Home   To move to the beginning of the cell contents

- End   To move to the end of a cell content

- Backspace   To delete a character from left

- Delete   To delete a character from the right

- Tab   To accept the autocomplete suggestion

- Ctrl + PageUp/Down + Arrows   For referencing a cell from another worksheet

Other Shortcuts

Keys: Functions:

- Ctrl + ;   To enter date

- Ctrl +:   To enter time

- Ctrl + '   To show formula

- Ctrl +]   To select an active cell

- Alt   To drive menu bar

- Alt + Tab   To open the next program

- Alt + =   To autosum

TIPS AND TRICKS

Select All With a Single Click

You may be familiar with the Ctrl + A shortcut for selecting all, but few are aware that all data may be selected in secs with just one click of a corner button, as described below.

Open Multiple Excel Files at Once

When you have numerous files to handle, there is a convenient approach to open them all in one click rather than opening them one after another. First, select whatever files you want to open and press the enter key on your keyboard; all files will open simultaneously.

Switch Back and Forth between Different Excel Sheets

When you have several open, it's quite inconvenient to switch between spreadsheets since working on the wrong sheet might completely derail a project. You can effortlessly switch to different files by pressing Ctrl + Tab. This function applies to other files and different Windows sessions in Firefox while running Windows 7.

Make a Brand-New Shortcut Menu

Save, Undo Writing, and Repeated Typing are the three most common shortcuts on the top menu. However, if you'd like to use many shortcuts, such as Cut and Copy, you can do so as follows. Add Cut and Copy again from left column to right column under File-> Options-> Quick Access Toolbar, then save. In addition, two more shortcuts have been added to the top menu.

Draw a Diagonal Line through a Cell

For example, you could need a diagonal connection in the first cell to divide distinct columns and rows while making a classmate's address book. The best way to make it is the following. Everybody is aware that Home-> Font-> Borders allows you to adjust a cell's borders and even add other colors. If you choose More Borders, though, you'll get more surprises, such as a vertical line. You may now make it right away by clicking it and saving it.

Add Multiple New Rows or Columns

You may know how to add a single new row and column, but if you need to add upwards of one of these, repeating this step X times is time-intensive. So instead, if you'd like to add X columns and rows above it or left, the best technique is dragging and choosing X columns and rows (X is two or more). Then, select Insert from the drop-down menu by right-clicking the highlighted rows and columns. Additional rows will be added above, and to the side of a column you selected before.

Copy and Move Data in Cells Quickly

If you'd like to quickly relocate one column with data in such a spreadsheet, select it and drag the cursor to the boundary. When the pointer transforms into a cross arrow icon, slide to move the column freely. What if you need to copy the information? Before dragging to relocate, hold down the Ctrl key; the column would copy all selected data.

Delete Blank Cells Quickly

For different reasons, certain default data would be blank. If you need to eliminate these to ensure accuracy, particularly when computing the overall average, the quickest method would be to filter everything out of empty cells and remove them all at once. Select the subject you would like to filter, navigate to Data-> Filter, undo Select All when the downward button appears, and then select Blanks as the last choice. All empty cells will appear right away. Return to Home and select Delete from the drop-down menu; all of them could be deleted.

Using a Wild Card in a Vague Search

You may understand how to activate the quick search by pressing Ctrl + F. However, two major wild cards were used during Excel spreadsheets to initiate a vague query; Question Mark and Asterisk. This is used once you are unsure about the desired outcome. One character is represented by a question mark, whereas an asterisk represents one or even more characters. What if you need to find a target result that includes a Question Mark with Asterisk? Don't forget to draw Wave Lines in front of everything.

Create a One-of-a-Kind Value in Such a Column

Although you are familiar with Filter's main purpose, few people know Extended Filter, which is frequently used when you're about to filter the distinct value from the data in such a column. Go to Data-> Advanced after selecting the column. There will be a pop-up window. Click Copy to another position, which should correspond to a second red rectangular region, as shown in the screenshot. Then type the value or click the area-choosing button to choose the target place. For example, column C can create the distinct age, which will be displayed in Column E in this case. Remember to choose unique records, then click OK. Because the unique value in column E may differ from the original information in column C, it's also suggested that you copy to a different location.

Data Validation Function with Input Restriction

To keep data legitimate, you may need to limit the data input and guide the next stages. For instance, age should be expressed in whole numbers on this form, and all survey participants should be between the ages of 18 and 60. To make sure data from outside this age bracket isn't inputted, go to Data-> Data Validation-> Setting; then set the conditions, and then switch to Embedding Process to give warnings like, "Please input your age as a whole figure, that will range from 18 to 60." When the pointer is hung in this location, the user will receive this prompt and a warning notice if the information entered is unqualified.

Ctrl + Arrow Button for Quick Navigation

You can jump to an edge of a sheet in various directions by pressing Ctrl + any direction key on the keyboard. If you wish to jump to the bottom of the data, use Ctrl + downward.

Change the Order of Data in a Row to the Column

You'd use this functionality to transpose data for a better display; but, if you know when to use Paste's Transpose function, copying pasting all data would have been the last thing you'd have to do. Here's how to do it. Move your pointer to another blank position and copy the region you want to transpose. Go Then, go Home-> Paste-> Transpose; n

Completely Conceal Data

Almost all users know how to conceal data by right-clicking and selecting the Hidden function, but it can be identified if there is just a little data. The Format Cells method is the finest and simplest approach to hide data completely. First, select the area and then navigate the Home-> Font-> Open Formatting Cells-> Number Tab-> Custom-> Type ;;; -> When you click OK, all of the values throughout the area will be hidden, and you'll only be able to see them in the previewing area beside the Function button.

Tricks

Use CTRL+1 to Format Any Object Quickly

When you pick any item in Excel, a cell, chart, chart axis, or drawing object, and press Ctrl+1, the Properties dialogue for that object appears. This shortcut allows you to format whichever object you're dealing with quickly and easily.

Use the CTRL+G or f5 Keys to Create Range Names

If you're using range names (as we highly encourage) and wish to select a range with particular name references, hit Ctrl+G or even the F5 key to open the GoTo dialogue.

If a name is straightforward, you can select it from a list within this dialogue. However, Excel won't list it if it's unusual, so you'll have to write it in manually. Then press the OK button.

Use =Sum(And f3 in a Function with a Range Name

Let's say you wish to use the name of a range in a formula. You might want to total the Sales range, for example. Enter =sum(And then press the F3 key.

Excel opens the Paste Name dialogue when you do so. Simply select "Sales" from the drop-down menu, click OK in the window, and then insert the SUM function's closing ")" to finish the formula.

Use CTRL+a to Open the Function Arguments Dialogue Quickly

Let's say you wish to look up a worksheet function's help topic. You might be interested in learning more about the MATCH function, for example. Fill in the blanks in a cell using =match(. Then, hit Ctrl+A or select The Insert Feature ("Fx ") button on The Formula Bar's Left.

Excel then displays its Function Arguments dialogue, which may provide all of the assistance you require.

However, if you still want to see the entire help topic, go to the lower-left corner of the window and click on the blue "Help on this function" hyperlink. This method works with all Excel functions that are documented.

Use CTRL+D to Copy Items Down the Column While Scrolling

If you wish to replicate a formula from a column on the right side of a large dataset without scrolling, follow these steps:

- Travel to a right to the data-filled column (the column to a left of a new formula-filled column);

- To reach to the bottom, press Ctrl+Down;

- Drag one cell to the right (Naturally using the arrow key);

- Ctrl+Shift+Up selects the new column with the formula you have created at the top; Ctrl+D fills the formula down.

With Alt+, You Can Quickly Access Any Function

You may build simple shortcuts to actions that you would then have to locate throughout the Ribbon tabs, even macros you have built yourself, by configuring the fast access toolbar.

Selecting Alt+ is the keyboard shortcut (the number of the command you wish to select).

For example, suppose you have Calc Sheet, Save, and Open on your fast access toolbar. To calculate a sheet, press Alt+1, save with Alt+2, then open with Alt+3.

Many users are ignorant of this valuable feature, which can save you a lot of time.

Use CTRL+1 to Format Cells

Use Ctrl+1 to format cells when necessary. Most people know this as a shortcut for the Format Cells dialogue, but you can format nearly anything in Excel without worrying about the ribbon's state. Try out this fantastic and easy shortcut!

Use Alt+ to Select Visible Cells

When you're about to select only visible cells, use Alt+. This is how you can only duplicate what you see. When manually hiding columns and rows in a table, this is a priceless shortcut.

Make Use of Filtering

Filtering is a powerful tool for slicing, dicing, and sorting a large table of data.

When you're in a group to decide something more like a sales forecast, and everybody is looking at your spreadsheet displayed on a screen in real-time, it's quite effective (or on their monitors).

Some individuals will know you as the King of Spreadsheets, but this is not a hoax!

With the CTRL Key, You Can Quickly Insert or Delete a Column or Row

Even simple activities, such as inserting and deleting rows and columns in Excel, might take a long time for certain people.

- To insert, hold down Ctrl+Shift ++ while selecting a whole row or column.

- To delete a full row or column, press Ctrl + – while selecting it.

Use f9 to Get the Formula Results

If you wish to examine the outcomes of various formulas, pick the formula and press F9 to display the result.

Before exiting the formula, remember to undo it.

To Add Extra Text to a Cell Press Alt+Enter

Use ALT+Enter to insert the second bit of text within such a cell.

To Advance a Date by Either a Full Calendar Month Use EDATE

Here's how to put EDATE to work for you:

- 15/02/2016 =EDATE(15/01/16,+1) (15th Feb 2016)

- 15/11/2015 =EDATE (15/01/2016,-2) (15th Nov 2016)

To Change a Date to the End of the Month, Use EOMONTH

Here's how to put EMONTH to work for you:

- =EOMONTH(15/01/2021,0)   =   31/01/2021   EOMONTH(15/01/2021,0) EOMONTH(15/01/2021,0) (31st Jan 2121)

- 30/11/2020 = EOMONTH (15/01/2021,-2) (30th Nov 2020)

LANGUAGE

For data processing, Excel files are commonly used. In certain cases, the program is very much needed to complete the study. This requires the use of a programming language that can easily parse Excel data.

Languages that may process and parse Excel files involve:

1. es Proc S.P.L.

2. Python;

3. Excel VBA;

4. High-level General-purpose programming languages, like Java;

What is Excel VBA

VBA is the abbreviation for Visual Basic for Applications. It is a programming language that can be used to automate certain tasks in Excel. VBA is the scripting language made by Microsoft. Many of the Office suite applications are written in the same programming language.

Why Use Excel VBA

Although VBA does not allow users to explicitly control the key Excel program, it does allow them to learn the art of creating macros in order to save time in Excel. Excel macros can be generated in two ways.

The Macro Recorder is the first approach. Excel can log all of the actions a consumer takes after triggering the recorder and store them as a "process" known as a macro. When the consumer closes the recorder, the macro is saved and can be assigned to a button that will repeat the operation when pressed. This approach is straightforward and does not need any prior knowledge of VBA technology. For basic operations, this approach would suffice.

The downside is that it is not quite flexible, and the macro would exactly duplicate the user's input. Recorder macros utilize absolute referencing instead of relative referencing by nature. It indicates that macros created in this manner are difficult to use for variables and "wise" methods.

The second and more efficient way to make an Excel macro is to use VBA code.

Where to Code Excel VBA

Inside any Office software, click Alt + F11 to open the VBA window. When performed correctly, on the top left, this will open a window with a file structure, while on the bottom left, a properties window appears, a debug pane in the bottom middle and bottom right, and the coding area in the center and top right, which will take up the majority of the screen. This may appear troublesome to start with, but in reality, it is less problematic than one might think.

The consumer will spend the majority of their time in the coding area. The file structure portion is only used while a new macro file is being created. Only more complex macros that use UserForms to build graphical environments for the macro can use the properties portion in the bottom left.

The majority, if not all, of the coding takes place in the coding section. This is where the user can build, code, and save macros. The macro code will then be added to specific triggers in the Excel model after it has been written and saved. The macro may be started by pressing a certain button on the worksheet or by changing the contents of some cells, for example. Attaching a macro to a button is the simplest way to do it.

High-Level Languages (Take Java as an Example)

Excel files can be interpreted in almost all high-end languages. The argument is that whether there isn't a skilled data retrieval A.P.I available, you'll have to compose your own software to read the data according to the format of the target Excel file. This needs a significant amount of time and commitment. Fortunately, Java has an Apache P.O.I., which allows you to read & write Excel data. Each cell's values & properties can be read by the API. Let's have a peek at how it reads Excel spreadsheets and converts them to organized files.

The software can only be interpreted in the simplest Excel file format. It's very lengthy, even though the handling of cell values isn't included. If an Excel file contains a complex structure, such as combined cells, multi-row records, complicated table footers and headers (multi-row), and crosstab, it can take longer and be more complicated.

Also, with such a strong open-sources package as the P.O.I., parsing Excel files in Java is still very difficult.

Furthermore, high-level language only has low-levels functions & special neglect functions for organized data computations such as filtering, grouping, sorting, and aggregation, and data set joins. Programmers could compose theirs in a unique way. And after the data has been read & parsed, there is already a lot of analysis to be done.

Python

The Python pandas have an Excel file reading G.U.I.

When the header parameter is set to 0 (the header = 0), the first row is interpreted as column headers. The collected, organized data collection is referred to as "data."

e.g.

	A	B	C	D	E	F	G
1			Item Lists And Prices				
2	Project:	Building				page 1/total 52	
3	No	Item Code	Item Name	Unit	Quanti	SumOfMoney(yuan)	
4					ty	Price	Sum
5	1.1.2	NJSJ	Internal scaffolding	term			
6	1.1.2.1	11001004001	Internal scaffolding	term	1.00	1006577.54	1006577.54
7	1.1.2.1.1.1	A22-28	Steel pipe	100m²	137.88	912.07	125756.21
8	1.1.2.1.1.2	A22-28	Base of internal scaffolding	100m²	71.83	912.07	65513.99

The software states that table headers are not to be read & that reading should miss the first four rows in order to start with the five throws through the parameters (If there are any table footers, users can specify that they should be skipped). In the last lines of the code, it creates an alias for the data collection "data." e.g.,

	A	B	C	D	E	F	G	H
1	Orders Statistics							
2	Type Area Amount	West	East	Center	North	South	Northwest	Southwest
3	Urgent express	20	70	1	97	23	2	35
4	Unified parcel	25	89	1	148	39	3	27
5	Federal cargo	15	79	52	108	29	2	23
6	Air transport	5	1	12	1	1	9	6
7	Cash on Delivery	8	2	4	1	6	7	9
8	General express	32	41	36	48	26	22	18

In Python, reading Excel files is easier than in Java. Python has better protection for subsequent computation than VBA and Java since Pandas encapsulates organized computing functions. This renders it simple for Python to be processing tiny Excel files which can fit entirely in memory.

The issue with Python's that it lacks batch processing ways for large Excel files, which can't be loaded into memory all at once. As a result, you'll need to write a complex program to read the data and perform corresponding calculations.

The esProc S.P.L.

The esProc, as a technical data management platform, has a number of options for reading Excel reports. S.P.L., the scripting language on which esProc is built, contains a large library of standardized computation functions which can manage all the subsequent computations as well as the exporting and writing of result sets to the database.

The esProc Excel data reading software is remarkably basic. There is just one section of JavaScript in it.

In parsing the Excel text, esProc SPL seems to yield more concise code than the Python pandas. But in reality, SPL has a competitive advantage when it comes to managing subsequent computations.

Furthermore, esProc SPL has a cursor mechanism that makes reading and computing large Excel files a breeze. This allows the data analyst to process the vast volume of the data using syntax that is close to that used to process small quantities of data, yet in a clear, intuitive manner.

PROBLEMS AND THEIR SOLUTIONS

Troubleshooting MS Excel 2021. Most individuals are either low on time and have a deadline to reach while dealing with Excel spreadsheets, making it a disaster if Microsoft Excel refuses to launch at any stage.

Unfortunately, when Excel declines to open one of the essential Excel files or refuses to open the Excel program at all, this behavior is very normal. So, if you're experiencing the dreaded "Microsoft Excel won't open" issue.

Here is a guide that will go through the explanations of why Excel won't open and what you should do about it.

Problem Opening MS Excel

There are aspects that may go wrong in Excel as it is a platform that must function together with other software and operating systems. Of course, it's also possible that the Excel program is the source of the issue.

The following are several potential causes for Microsoft Excel not opening on your computer:

- You are unable to open Excel files due to a malfunctioning add-in. You would be able to open Excel programs but not individual files or new ones.

- There's a chance the Excel program is corrupt, and you'll need to patch it before you can access Excel files again.

- The Excel program is unable to interact with other programs or the operating system. This is an easy fix; simply allow the configuration, and your files can begin to function again.

- The file association is disabled, which ensures it doesn't know what program to use to access an Excel file when you want to open it. You can easily correct this by merely resetting the file associations. This is a common issue with those who have updated Microsoft Excel or Microsoft Windows.

- The file you're trying to access is corrupted. This also occurs when you share a file with someone else or copy a file to an external drive or to a network drive and then move it to your device. There's not anything you can do about this situation except make sure you back up your data in the future.

Excel Files That Won't Open

Let's have a peek at some of the approaches for resolving the problem of Excel files not opening. These corrections are listed in the order in which they are most likely to perform.

Uncheck the box that says "Ignore DDE."

The most popular and simplest solution is to ensure that the proper setting for Dynamic Data Exchange is allowed (DDE). DDE is the method by which Excel communicates with other programs.

The DDE setting is disabled by default, but if it is allowed by mistake, it can prevent your Excel files from opening on your device. You'll be able to access the Excel program from the start menu, but you won't be able to open specific Excel files if this occurs.

The measures to resolve the DDE problem are as follows:

- From the Start menu, choose the Excel file.

- Go to the File Tab.

- Choose Options.

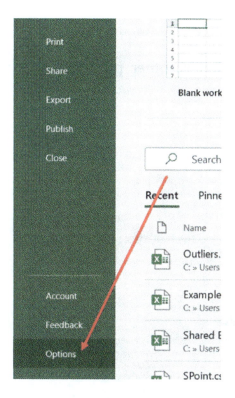

- In the Excel Options dialogue box that appears on the left pane, press the "Advanced" button.

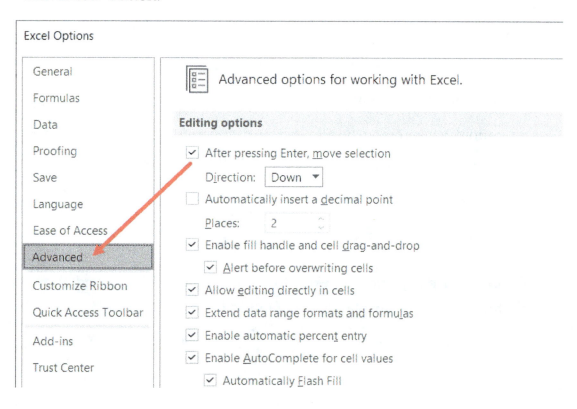

- Go to the General portion and scroll down.

- Disable the option to ignore other programs that use Dynamic Data Exchange (DDE).

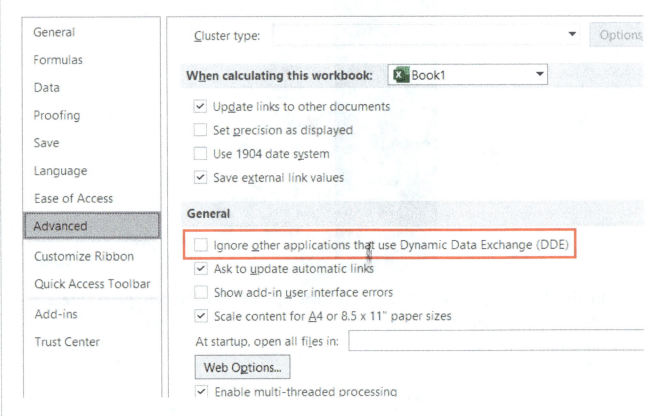

- Exit the Excel Options window.

If this configuration was previously allowed and you removed it using the measures above, this was most definitely the issue, which should now be resolved.

Continue reading if this approach doesn't fix your issue.

How to Turn Off an Add-in

Many people use third-party Excel add-ons to extend the capabilities and allow them to use several features that aren't available by default in Excel.

One of the add-ins is "ThinkCell," which helps to make beautiful graphs and charts that are not achievable with Excel alone. These add-ons can also be the cause of you being unable to access your Excel files. Disabling the add-in is the simple solution.

The following are the measures to uninstall an add-on in Excel:

1.   From the Start menu, choose the Excel file.

2.   Go to the File Tab.

3.   Select Options.

4.   Hit the Add-ins icon in the left pane of the Excel Options dialogue window.

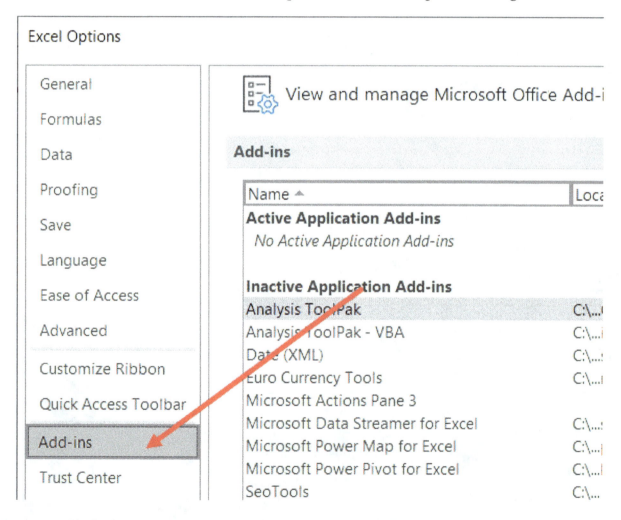

5.   Click the Manage menu at the base of the dialogue box.

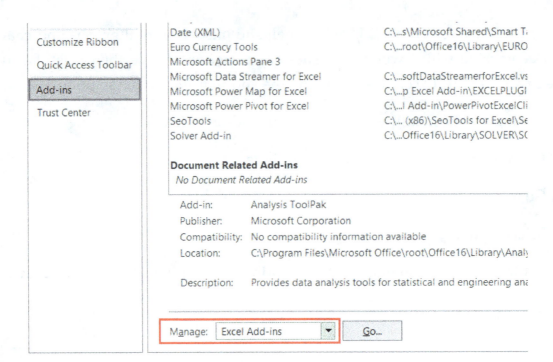

Customize Ribbon	Date (XML)
	Euro Currency Tools
Quick Access Toolbar	Microsoft Actions Pane 3
	Microsoft Data Streamer for Excel
Add-ins	Microsoft Power Map for Excel
	Microsoft Power Pivot for Excel
Trust Center	SeoTools
	Solver Add-in

C:\...s\Microsoft Shared\Smart T.
C:\...root\Office16\Library\EURO

C:\...softDataStreamerforExcel.v:
C:\...p Excel Add-in\EXCELPLUGI
C:\...l Add-in\PowerPivotExcelCli
C:\... (x86)\SeoTools for Excel\Se
C:\...Office16\Library\SOLVER\S(

Document Related Add-ins
No Document Related Add-ins

Add-in:	Analysis ToolPak
Publisher:	Microsoft Corporation
Compatibility:	No compatibility information available
Location:	C:\Program Files\Microsoft Office\root\Office16\Library\Analy
Description:	Provides data analysis tools for statistical and engineering ana

Manage: Excel Add-ins   ▼   Go...

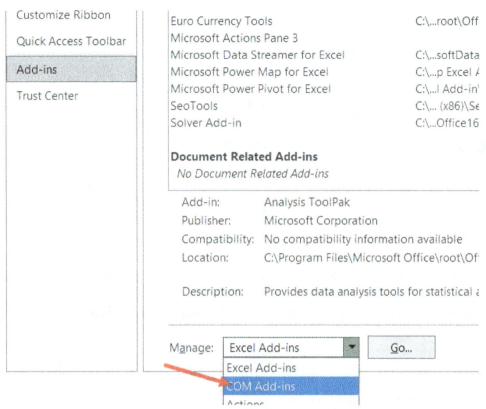

6. Select COM Add-ins from the drop-down menu.

7. Go by pressing the Enter key.

8. Disable all add-ins in the COM Add-ins dialogue box that appears.

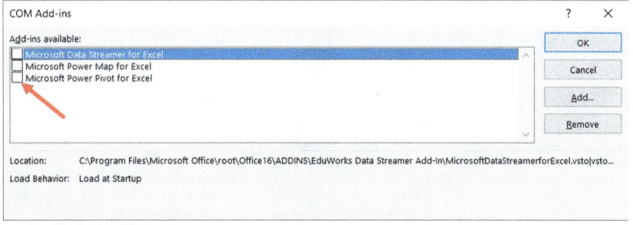

9. Click the OK button.

Now try to access an Excel file that wouldn't open previously. If this was the issue, then the file should open normally.

Often, it's a one-time thing, and the add-in can continue to function normally if you activate it again. If you allow the add-in and the issue persists, it is likely that the add-in is corrupt and has to be permanently removed.

Microsoft Office Needs to Be Repaired

Another explanation Excel files aren't opening is because the Microsoft Office program is corrupt and needs to be restored or reinstalled. But, as reinstalling is a bit more effort, let's start with the repair alternative.

To restore the Microsoft Office program on your device, follow the measures below:

1. Press the R key when holding down the Windows key. The "Run" dialogue box will appear.

2. Type the command "appwiz.cpl" into the run box. This will cause the dialogue box to pop-up for programs and tools.

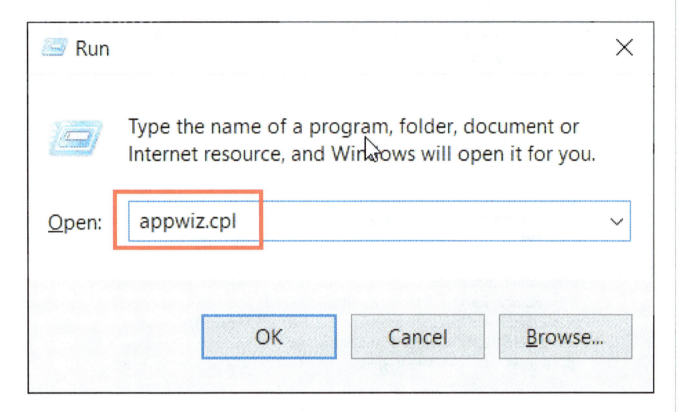

3. Choose Microsoft Office from the drop-down menu.

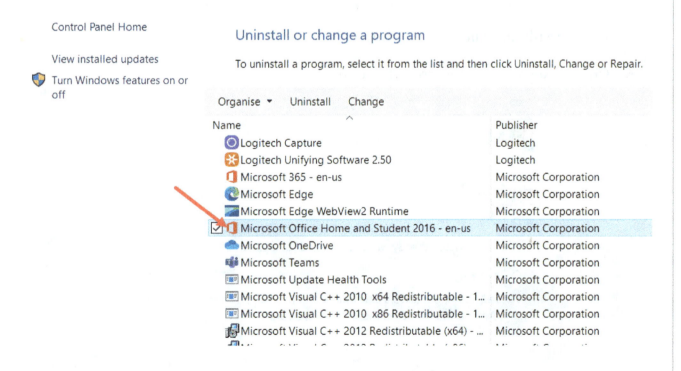

4. Change the Microsoft Office option by right-clicking on it. (If you see the button to repair right here, choose it.)

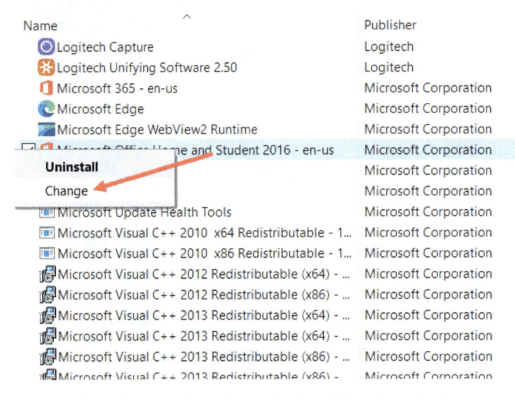

5. In the resulting dialogue box, choose the Quick Repair method.

6. Choose Repair from the drop-down menu.

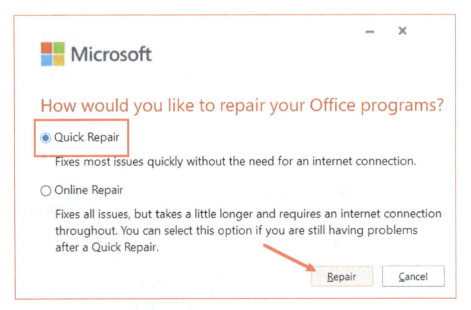

Follow the on-screen instructions to repair your Microsoft Office program in a matter of minutes. If the problem was caused by a compromised Microsoft Office program, following the measures above will resolve it.

Let's look at some more fixes if you're still having trouble opening Excel data.

Excel File Associations Can Be Reset

When you launch an Excel file, the file association ensures that the Excel application is used to open it. And, occasionally, these file associations go wrong, so when you press on an Excel file, it doesn't recognize that it needs to be opened with the Excel application. Resetting file associations is the solution.

The instructions to achieve this are as follows:

1. To begin, open the Control Panel.

2. Go to Programs and select it.

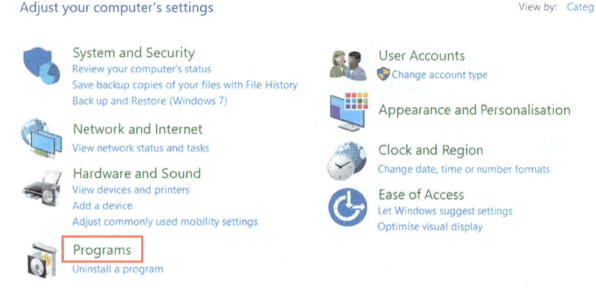

3. Select "Default Programs" from the drop-down menu.

Choose the programs that Windows uses by default

 Set your default programs

Make a program the default for all file types and protocols that it can open.

 Associate a file type or protocol with a program

Make a file type or protocol always open in a specific program.

 Change AutoPlay settings

Play CDs or other media automatically

 Set program access and computer defaults

Control access to certain programs and set defaults for this computer.

4. Select "Set your default programs" from the drop-down menu.

 Programs and Features
Uninstall a program | Turn Windows features on or off | View installed updates |
Run programs made for previous versions of Windows | How to install a program

 Default Programs
Change default settings for media or devices

5. Scroll to the "Reset to the Microsoft recommended defaults" section of the Default apps window that appears, then click the Reset button.

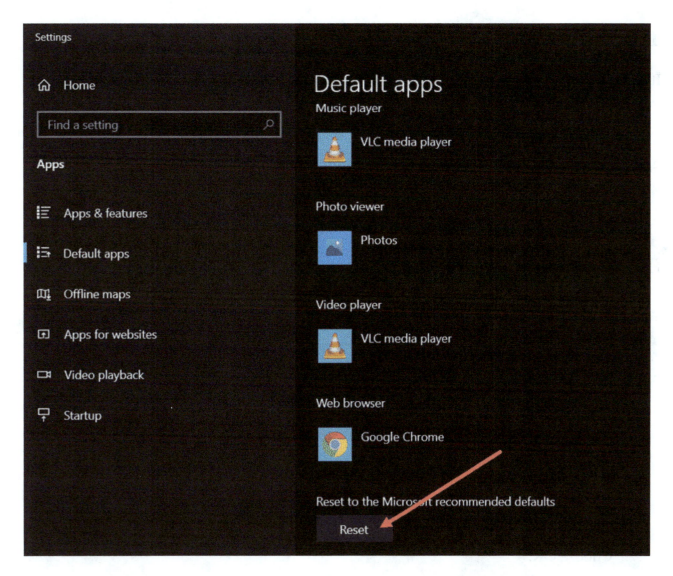

If a mismatched file association was preventing your Excel files from opening, this would solve the problem.

Because we've reset the default for all the applications on your device, any manual changes you've made, such as specifying a particular file extension to be launched with a particular application, will be reset to the default as well.

If you don't want to reset the default for all applications but only for Excel files, click the "Choose default applications by file type" option.

Find the Excel file extensions (.xls,.csv,.xlsx,.xlsb, etc.) and make XLS the default software for those on the screen that appears.

Turning off Hardware Graphic Acceleration

It improves the performance of your device, particularly with Microsoft Products like MS Word or MS Excel. However, this will sometimes result in your Excel files not working or cause your device or files to crash.

So, if everything else fails, this strategy is worth a shot.

The problem can be resolved by removing hardware graphic amplification, which is allowed by design.

Here are the steps to follow:

1. Launch the Excel programs.

2. Choose the File page.

3. Choose Options.

4. In the Excel Options dialogue box that appears, in the left pane, press the Advanced button.

5. Go to the "Display" settings and scroll down.

6. Choose "Display hardware graphic acceleration" from the drop-down menu.

7. Choose OK.

If hardware graphic acceleration was the issue, your Excel files should now open.

There are many things you can do in terms of troubleshooting. If nothing else seems to be working, you can contact Microsoft support to see if they can assist you.

More often than not, these problems can be traced back to a Microsoft update that was published recently, which unwittingly triggered the issue.

A further option is to post your question on one of the several active Microsoft forums, where the wonderful people of the Internet can assist you.

APPLICATIONS AND BENEFITS IN AN ORGANIZATION

To remain competitive in today's world, all businesses must grow and progress. Implementing development plans so that workers can stay on top of current technology and function as effectively as possible is one way to stay ahead of the pack and encourage profitability.

Talented employees want to be tested and work hard to stay ahead of the competition. Employers can increase productivity and lower employee turnover by offering them the education and training they need to be as successful as they want to be. They can also reduce the risk of losing the most skilled employees to rivals by facilitating them with the ongoing training they need to be as effective as they want to be. Excel for business is a curriculum that is often used in these education training programs.

What Is Excel Used For

Excel allows users to measure, organize, and interpret quantitative data, enabling managers and senior staff to make strategic decisions that can impact the organization with the information they need. Employees who are qualified in specialized Excel functions would be able to present their data more effectively to upper management. It's also a necessary skill for workers who want to work their way to the top.

Employees and employers would both benefit from advanced Excel knowledge. Let us now examine the benefits of Excel when it's part of a company's regular employee preparation.

Excel's Benefits for Employees

Advanced Excel training can help employees in a variety of ways, from increasing their value to learning new resources to increase their job performance.

Making Improvements to Your Skill Set

To advance your career, you must continue to learn and improve your skills. Advanced Excel training covers a variety of important skills that can be applied and appreciated in almost every job position. You should be able to:

- Visualize, manipulate, and assess data after practicing.

- Develop equations that will enable you to have more information on critical company functions like workflow, project performance, financial forecasts and budgets, and perhaps even inventory levels and utilization.

- Create an easy-to-understand data set that upper management may use to assess current tasks or conditions in the organization.

- Create spreadsheets that help organize data and provide a clearer image of what's being entered.

- Read and understand data and spreadsheets from other agencies, suppliers, and consumers.

- The ability to view data at a higher level allows you to have answers and solutions to business problems.

- Plan, balance, and maintain complex financial and inventory records.

- Set up monitoring systems for various departments and operations, as well as different workflow processes.

- Advanced Microsoft Excel preparation will also provide employers with higher-skilled workers as well as opportunities to help employees work more efficiently in their current positions and prepare them for advancement to higher-level positions.

Increasing Your Productivity and Efficiency

When working with large quantities of data and measurements, Excel is a critical tool for improving productivity and increasing efficiency. When you develop a greater understanding of Excel, you will be able to use its more advanced software, which will help you to complete tasks and analyze data more quickly. It will also enable you to keep members of the team informed about data, which will help to speed up the workflow process.

Even better, learning advanced Excel will help you improve the efficiency of your calculations. Calculations that have to be repeated take time, particularly when you have to dual-check your work. You can make more complicated calculations using advanced Excel software. If you've written your formula and programmed your preset command, the software will complete the calculations for you, freeing up time for other activities and ensuring that you get accurate results the first time.

Making Yourself a More Valuable Company Employee

Being a productive member not only ensures your job security but also allows you to progress your career. Being more effective, better trained, and skilled in your work activities would help you become more valuable to the business. That is what advanced Excel training should provide. To avoid being replaced by younger staff with a more sophisticated skillset, employees should always look for further ways to optimize. To remain on top of the game and set yourself up for enhanced stability and progression, you must learn and master new skills.

Making You Better at Organizing Data

Spreadsheets are a popular tool for collecting and organizing data. Excel is the spreadsheet software in its most basic form. It helps you to meticulously organize all of your data while still allowing you to sort the data in whatever way you want. Data in its raw form can be confusing and difficult to understand. With Excel's advanced features, you'll be able to organize your data better, perform calculations as required, and sort the data so it can be properly analyzed and transferred to graphs or charts for better viewing.

It Can Make Your Job Easier

The great your familiarity with Excel, you will be able to use the system. Microsoft Excel has a number of shortcuts that can help you work quicker and also learn more complex

Excel techniques that you can apply to the entire Microsoft Office suite. You'll also be able to access the information in Excel sheets in a number of applications, reducing the need to re-enter data and improving the efficiency of your workflow. The simpler your job is to do and the more prepared you are to do it; it is more probable that you will like it. Indeed, studies have shown that happier employees are 20% more efficient than their unhappy colleagues. You will be a happier and more efficient employee if your job is easier.

Advantages of Advanced Excel for Employers

Advanced Excel training and expertise will provide multiple benefits for both your employees and your company. It boosts productivity and increases performance. Yes, as mentioned, advanced Microsoft Excel training will enhance employee efficiency and productivity, resulting in increased efficiency and productivity for the organization. The more productive your staff work, the faster tasks and projects can be completed, allowing you to offer better service to your customers and partners while still producing more work in a shorter amount of time. Even if the advantages of advanced Excel training save your employee a half-hour per week, when compounded by the number of workers in the department or business, it can add up to a large number of additional staff hours per week.

It gives you the opportunity to use a positive situation that you already have better.

Your company's software systems are assets, and if your workers aren't qualified to make the most of them, they could be considered underutilized. Continued Excel training will enable you to get the most out of that asset, as well as other resources that may not be used to their full potential, like inventory management systems. For instance, if your workers can properly coordinate and streamline calculations, you can boost inventory control and make better use of your assets.

It enables you to increase employee knowledge with minimal cost and effort. Workers in your company have already been trained in the original Excel software and introducing basic training programs that help you to properly use the program can be far less costly than trying to train new workers in your company's processes and procedures, who already have experience of these advanced systems. Furthermore, advanced training can be simple, taking just a few weeks or less for workers who have already demonstrated intermediate proficiency. Rather than investing in outside training programs for and employees, you can save money by employing an onsite trainer who can train a huge group of employees at once. It results in a more trained and professional workforce at a lower rate.

Excel Improves the Operation of the IT Support Team

It's up to the IT department to take up the slack when workers aren't well educated in all facets of a software program. IT staffers moving from workstation to workstation to train coworkers prevents them from concentrating on more productive tasks, including system updates, protection, and hardware installation and maintenance.

Furthermore, just because the IT department will assist employees with software use does not imply that this is employed for a specific goal, and they are providing the necessary data and information for the goods. Their skills are more specialized, and they may not know the significance or role of the data they are assisting the employee with. If the workers are qualified in advanced Excel, they would be able to manage their own data manipulations, saving time and delivering better results than if they had to rely on IT.

It will aid your talent retention efforts and provide a more satisfying work environment for your employees.

Employees who are valued thrive on acquiring new skills that will allow them to not only succeed in their current role but also advance up the corporate ladder. If you don't feed this desire to learn, your employee's job satisfaction will suffer, and they will be less motivated to continue their career path with your business. Employees become more important when they are trained. When you educate workers, you increase their value to the company while also lowering turnover and giving the best employees a reason to stay.

Employee preparation is an important part of advancing the workforce, increasing efficiency, and maintaining a strong workforce. Whether you want to invest in onsite training to keep your employees up to date on advanced Excel operations or enable them to seek outside training opportunities, such as a master's program with Advanced Excel coursework, continuous learning for your employees is critical to help you develop your business and stay ahead of your industry's competitors.

Awareness is a strength, as they claim, and there's no better way to inspire the employees, develop their skills, and increase their value to the organization than encouraging them to use critical programs to their full potential. Using advanced Excel training to enhance the employee's day-to-day work will keep them engaged, improving, and delivering efficiently.

CONCLUSION

The new Excel models provide all you need to get started and become a specialist, including a variety of great functionality. MS Excel recognizes patterns and organizes data to save you time. Create spreadsheets easily from templates or scratch, then perform calculations with modern features.

It's built with both simple and advanced applications that can be used in almost every business environment. The Excel database allows you to create, view, edit, and share data with others quickly and easily. When reading and editing excel files attached to emails, you will create spreadsheets, data tables, data reports, budgets, and more. As you've become more acquainted with different concepts, you can know the new resources and functionality that Excel provides to its users. The truth is that with Excel features, you can meet almost every person or company requirement. What you need to do is invest your time and expand your knowledge. The learning curve for improving your skills will be long, but you may find that items become second nature with experience and time. After all, a man becomes better through practice.

Everything you must do to make your life easier—and potentially impress others in your office—is to master these fundamental Excel skills. But Don't neglect that no matter how experienced you are with this versatile tool, there is still something new to discover. Whatever you do, continue to keep improving your Excel skills—it will help you not only keep track of your own money, but it may also contribute to a better career opportunity in the future.

Microsoft excel 2021, the new version, is the continuation of v6, with a new version that adds new features. This software is used to view files, manipulate data and work with numbers, dates, and texts. Microsoft excel 2021 can be installed on your system without any cost. This software has a lot of features, such as easy-to-use functions. Microsoft excel 6 has many versions, which were released in the year 2004 to 2021. V6 exam is the latest version, and this software is compatible with Windows 10.V6 is the easiest version to work with. It is used to manage and organize work and personal documents, which helps to reduce the use of unnecessary paper documents.

This version is available in two forms—personal and enterprise. You can purchase the software online as well as download it. It is available in many languages and for several users. You can also use this software as a desktop application as well. If you are a beginner at Microsoft Excel, you have to perform certain actions. To start using the software, you need to have a trial version. Once you use the trial version, you can use the software and try the functions which are offered by it. You can likewise upgrade to the latest version as you have to with all the latest software available online.

If you're subscribed to Microsoft Office, you can use the latest version of the software. You can also use this application to discover more about the features it provides to the user. If you don't like the earlier version, you can get the latest version and use it. This version is easy to download as you need to visit the site to download it. There are thousands of users who can use this software to get the latest features and updates. You can even learn to use the features and learn more about this software.

www.ingramcontent.com/pod-product-compliance
Lightning Source LLC
Chambersburg PA
CBHW082119070326
40690CB00049B/3990